Signe Lin Kuei Vehusheia

Exploring heat flux as a marker for the early detection of infection

Signe Lin Kuei Vehusheia

Exploring heat flux as a marker for the early detection of infection

Hartung-Gorre Verlag Konstanz

Reprint of Diss. ETH No. 30190

Scientific Reports on Micro and Nanosystems Volume 39

edited by Prof. Dr. Christofer Hierold
ETH Zürich
Micro and Nanosystems

Cover images:
Front:

- **Left:** Image of *E. coli* MG 1655 bacteria
- **Middle:** Microfluidic chip used for chemical exothermic experiment and bacterial experiment
- **Right:** Image of THP-1 monocyte used for immune cell response experiments

Back:

- **Left:** COMSOL simulated heat transfer coefficient at the aortic wall with aortic pulsatile pressure-driven flow
- **Middle:** Top view of the microfluidic chip used for chemical exothermic experiment and bacterial experiment
- **Right:** Heat flow measured from the exponential growth of bacteria in the microfluidic chip

Bibliographic information published by Die Deutsche Nationalbibliothek

Die Deutsche Nationalbibliothek lists this publication in the Deutsche Nationalbibliografie; detailed bibliographic data is available in the internet at http://dnb.dnb.de.

First edition 2025
HARTUNG-GORRE VERLAG, KONSTANZ
http://www.hartung-gorre.de

ISSN 2566-7769
ISBN 978-3-86628-835-5

Diss. ETH No. 30190

Exploring heat flux as a marker for the early detection of infection

A thesis submitted to attain the degree of

DOCTOR OF SCIENCES
(Dr. sc. ETH Zurich)

presented by

SIGNE LIN KUEI VEHUSHEIA

M.Sc. in Material Science, ETH Zurich

born on 06.05.1995

accepted on the recommendation of
Prof. Dr. Christofer Hierold
Prof. Dr. Roland Zengerle
Prof. Dr. Goran Stemme
Dr. Nikola Cesarovic

2024

Abstract

This thesis investigates the feasibility of employing heat flux sensors for the early detection of implant infection. The model system on which this hypothesis is investigated is the aortic vascular graft. The investigation is motivated by persisting high mortality rates of patients with implant infections. In addition, the number of patients receiving treatments involving implants increases every year. Infections are hard to detect within the window of effective antibiotic treatment, which is at the early stages of infectious growth.

Many chemical and biochemical approaches have been investigated in literature as markers for implant infections. However, these sensing mechanisms are mostly used *in vitro*, as they are limited by drift due to foreign body reaction *in vivo*. Although known for centuries (1st century CE) that infections lead to local temperature increases, the use of physical sensors, especially heat flux sensors, is a lesser explored avenue. Using heat flux sensors for the detection of localized heat production by infections shows potential for long-term infection detection on implants. This sensing modality should be less affected by the foreign body response (encapsulation) as it is independent of analyte interaction with the sensor surface. Due to the principle of heat flux sensing it is also less susceptible to environmental temperature variations.

By means of *in vitro* microfluidic calorimetric chip experiments, the possibility of early detection of infections on aortic vascular grafts using heat flux as a marker is explored. This chip relies on a differential compensation scheme to sense the heat flow produced solely by bacterial growth. The limit of detection of the measurable heat flow density *in vitro* is experimentally established as 1707 W/m^3, which corresponds to 0.5% of the heat flow density of a bacterial biofilm or 9.2×10^5 bacteria (*E. coli*). This limit of detection is lower than the initial seeding concentrations typically used in *in vivo* implant infection studies in literature. The results of the *in vitro* microfluidic chip experiment thereby support that heat flux sensors are capable of early detection of infection. Furthermore, the flexible fabrication process of the *in vitro* microfluidic chip allows integrating an additional top channel on the microfluidic structure to mimic aortic blood flow. This addition increases the heat transfer efficiency of the microfluidic chip by 20%, and thereby pushing the limit of detection potentially even lower.

Finite element modelling results of an aortic system with an integrated heat flux sensor in an implant show a heat flow density limit of detection of 2000 W/m^3. Thereby, the simulated results of an aortic system align well with the experimentally determined values of the *in vitro* microfluidic chip. The experimental and modelling results shown in this thesis strongly support the case that applying heat flux sensing is a viable approach for the early detection of infectious growth on implant surfaces.

Zusammenfassung

In dieser Arbeit wird die Machbarkeit des Einsatzes von Wärmeflusssensoren zur Früherkennung von Implantatinfektionen untersucht. Das Modellsystem, an dem diese Hypothese untersucht wird, ist das Aorten-Gefässtransplantat. Grund für diese Forschung ist die anhaltend hohe Sterblichkeitsrate von Patienten mit Implantatinfektionen und die jährlich steigende Zahl von Patienten, die deshalb behandelt werden müssen. Infektionen sind innerhalb des Zeitfensters einer wirksamen Antibiotikabehandlung, d. h. in den frühen Stadien des infektiösen Wachstums, bis jetzt nur schwer zu erkennen.

In der Literatur sind viele chemische und biochemische Ansätze als Marker für Implantatinfektionen untersucht worden. Diese Erkennungsmechanismen werden jedoch meist *in vitro* verwendet, da sie durch Drift aufgrund von Fremdkörperreaktionen *in vivo* begrenzt sind. Obwohl seit Jahrhunderten (1. Jh. n. Chr.) bekannt ist, dass Infektionen zu lokalen Temperaturerhöhungen führen, ist der Einsatz physikalischer Sensoren, insbesondere von Wärmeflusssensoren, ein noch wenig erforschter Ansatz. Der Einsatz von Wärmeflusssensoren zur Erkennung von lokaler Wärmeflussproduktion durch Infektionen zeigt Potenzial für die langfristige Erkennung von Infektionen an Implantaten. Diese Sensormodalität sollte weniger durch die Reaktion von Fremdkörpern (Verkapselung) beeinträchtigt werden, da sie unabhängig von der Wechselwirkung des Analyten mit der Sensoroberfläche ist. Aufgrund des Prinzips der Wärmestrommessung ist sie auch weniger anfällig für Temperaturschwankungen in der Umgebung.

In dieser Arbeit wird die Möglichkeit der Früherkennung von Infektionen auf Aorten-Gefaesstransplantaten durch Wärmefluss als Marker anhand von Experimenten auf einem mikrofluidischen kalorimetrischen Chip untersucht. Der Chip basiert auf einem differenziellen Kompensationsverfahren, um die ausschliesslich durch das Bakterienwachstum erzeugte Wärmefluss zu messen. Die Nachweisgrenze der messbaren Wärmeflussdichte *in vitro* wird experimentell auf 1707 W/m^3 bestimmt, was 0.5% der Wärmeflussdichte eines bakteriellen Biofilms oder 9.2×10^5 Bakterien (*E. coli*) entspricht. Die Nachweisgrenze ist niedriger als die in der Literatur üblicherweise in *in vivo* Implantatinfektionsstudien verwendeten Ausgangskeimkonzentrationen. Die Ergebnisse des *in vitro* Mikrofluidik-Chip-Experiments unterstützen somit, dass Wärmeflusssensoren in der Lage sind, eine Infektion frühzeitig zu erkennen.

Darüber hinaus ermöglicht der Herstellungsprozess des Mikrofluidik-Chips die Integration eines zusätzlichen Kanals oberhalb der mikrofluidischen Struktur, um den Blutfluss in der Aorta nachzuahmen. Dieser Zusatz erhöht die Wärmeübertragungseffizienz des mikrofluidischen Chips um 20%, wodurch die Nachweisgrenze möglicherweise noch weiterreduziert wird.

Die Ergebnisse der Finite-Elemente-Modellierung eines Aortensystems mit integriertem Wärmestromsensor in einem Implantat zeigen eine Nachweisgrenze für die Wärmeflussdichte von 2000 W/m^3. Dabei stimmen die simulierten Ergebnisse eines Aortensystems gut mit den experimentell ermittelten Werten des *in vitro* Mikrofluidik-Chips überein. Die in dieser Arbeit gezeigten experimentellen und modelltechnischen Ergebnisse unterstützen nachdrücklich die These, dass die Anwendung von Wärmeflusssensoren ein praktikabler Ansatz für die Früherkennung von infektiösem Wachstum auf Implantatoberflächen ist.

Acknowledgements

Most importantly, and therefore firstly, I would like to extend my deepest appreciation to my parents, Jan and Feng-Ling, and my grandmother whose constant support and guidance have been instrumental in shaping both my academic journey and personal growth.

To my father, Jan, your unwavering belief in my abilities and dedication to nurturing my mathematical and analytical skills have been invaluable. Your inspiration and encouragement have empowered me to pursue a career in engineering and given me the confidence to thrive in traditionally male-dominated environments. Your steadfast belief that gender should never be a barrier to anything one sets one's mind to has been a guiding principle, and I am profoundly grateful for your love and support.

To my mother, Feng-Ling, your tireless work ethic and commitment to excellence have been a constant source of inspiration. Your emphasis on perseverance and hard work which you instilled in me since I was a child has given me the resilience needed to overcome challenges and pursue my goals with determination. Your sacrifices and dedication to providing me with the best possible education, have laid the groundwork for my academic success. I am deeply thankful for your support and encouragement.

Together, you have provided me with the love, guidance, and support needed to grow into the person I am today navigating academia and pursuing my passions. Your sacrifices and belief in my abilities have been the driving force behind my achievements, and I am forever grateful for your endless love and support. I love you both.

To my grandmother, who despite her 90 years, has through her devotion and prayers been a constant presence, serving as a source of strength and encouragement throughout my doctoral journey. Her belief in me, and her daily prayers, have been my guiding light during both challenging and triumphant moments. Her continuous support speaks volumes, reminding me of the boundless love that surrounds me. My grandmother's role as my number one cheerleader has been instrumental in shaping the person I am today, and for that, I am forever thankful.

I would like to express my sincere gratitude to the following individuals and organizations for their academic support and assistance throughout my PhD journey:

I with to express my gratitude to Professor Christofer Hierold for his support, patience, and guidance throughout my doctoral journey. From the very beginning he believed in me and the project. Even though I was warned of the valley of tears in the middle of the PhD journey, I could not have gotten through it without his support and mentorship. His unwavering guidance during challenging times and academic guidance provided me with the tools and strength to overcome obstacles and continue my PhD journey. He has graciously given me the opportunity to explore various avenues, regardless of the outcome, fostering an environment where experimentation is valued and embraced, thus contributing significantly to my personal and academic growth and development. In addition to his support, his approaches to both intricate project management and scientific challenges will serve as a guide in my future endeavors. In the future, I will find myself contemplating: WWHD (What Would Hierold Do) before making critical decisions. In addition to his support, he has connected me with valuable contacts and provided guidance that has proven important in navigating academia. Without his encouragement and belief in my abilities during difficult times, I would not have been able to persevere and achieve my academic goals. His dedication to my success has left an indelible mark on my academic journey, and I am honored to have had the privilege of being his student.

I extend my sincere gratitude to my co-supervisors, Dr. Markus Arnoldini and Professor Viola Vogel, for their support and guidance throughout my doctoral journey.

Dr. Markus Arnoldini of the Laboratory of Mucosal Immunology has stood by my side in the laboratory and provided guidance and help as I as a greenhorn just started in biological research. His expertise, patience, and support have been instrumental in my growth in the biological laboratory. I am grateful for his time, dedication and teachings, and for the many countless ways he has contributed to my development. This has paved the way for multiple publications together which is a testament to his commitment to the project.

Professor Viola Vogel of the Applied Mechanobiology Laboratory has supported my project starting already from the very beginning by giving feedback regarding my research hypothesis and connecting me with Professor Emma Slack. Her endorsement and access to her laboratories and research group have been assets to my research in the cell culturing laboratory. Her reputation as a leading figure in the field has been a source of inspiration, and I am deeply thankful for her endorsement and support of my work.

I would like to further extend my gratitude to Professor Emma Slack, Dr. Olivier Braissant, and Dr. Nikola Cesarovic who each have helped out in the fields I had no previous experience in. They have introduced me to the bacterial growth in the biological laboratory, microcalorimetric measurements of cells, and given insights in cardiovascular diseases and the medical field, respectively. Without their help, this interdisciplinary work would not have come to fruition.

Professor Emma Slack of the Laboratory of Mucosal Immunology has graciously welcomed me in her research group and providing access to her biological laboratories and to her group. I am also deeply appreciative of the warm welcome and support extended to me also by the members of Professor Slack's research group. Their camaraderie and encouragement have fostered a collaborative environment where I have received so much help and always felt comfortable with asking questions. Their contributions have played a significant role in shaping my academic and personal growth, and I am privileged to have been a part of their group.

Dr. Olivier Braissant from the Department of Biomedical Engineering at the University of Basel has shared his vast knowledge in microcalorimetry and personally worked with me in the laboratory. I reached out to Olivier after reading an exceptionally informative and well-written review paper about microcalorimetry that he authored. I expressed my desire to meet and discuss my own project and through Olivier's welcoming nature and enthusiasm a fruitful collaboration evolved which has proven integral for many of the results in this thesis. He joined the project completely out of interest, and has taught me the ropes of different microcalorimetric measurements. I am grateful for the willingness to teach and genuine interest. This fruitful collaboration has resulted in multiple publications which are integral to my PhD and I am grateful to have had the opportunity to work alongside him.

Dr. Nikola Cesarovic of the Translational Cardiovascular Technologies at ETHZ and of the German Heart Center Berlin is both a coauthor and an examiner. Throughout my PhD he has provided valuable insights and feedback that have enriched my work. Specifically, his expertise in the medical field has been particularly valuable, serving as a crucial resource for medical input regarding my project which has been greatly appreciated. This ranges from the early-stage brainstorming sessions to later-stage specific medical questions. Further, I am grateful to Nikola for facilitating connections with Petra and Tine, whose assistance in the lab has been integral to the progress of my research. His involvement has important to the success of my thesis, and I am honored to have had the opportunity to coauthor papers wit him in addition to have had the opportunity to learn from him.

I would like to highlight Christien Beez (Tine) from the German Heart Center Berlin and Petra Wolint from the Translational Cardiovascular Technologies at ETHZ for their tireless help regarding cell culturing. Tine, based in Berlin, has been instrumental in helping with remotely regarding protocols for immune cell culturing. Her knowledge and guidance have been extremely helpful in learning about immunology research which I had no prior familiarity with. I am also happy that we got to meet and discuss in person one afternoon in Berlin. Petra has been a steadfast mentor and support system in the laboratory, tirelessly teaching me all of the different protocols of cell culturing in the laboratory. Her assistance and willingness to lend a helping hand whenever I had questions in the cell culturing laboratory have been exceptional. Petra's dedication and generosity exemplify the spirit of teamwork and camaraderie that defines our research environment. Both of their contributions have been indispensable, and I am honored to have had the opportunity to collaborate with them.

I would like to express my gratitude for Professor Volkmar Falk the Medical Director and Director of the Department of Cardiothoracic and Vascular Surgery of the German Heart Center Berlin whose endorsement of my PhD has streamlined the collaborations with Nikola, Tine, and Petra. Further his interest, encouragement and endorsement of my project already at the very beginning of my PhD has been inspiring and of utmost help. Despite his demanding schedule, he has offered his time to discuss my project on multiple occasions. His shared insights of the field of cardiovascular implant research have been enlightening. Furthermore, the tour he organized for me of the surgical rooms at the German Heart Center of Berlin made a lasting impression.

I would also like to extend my appreciation to my further committee members, Roland Zengerle and Goran Stemme for taking the time out of their busy schedules to read through and be a part of my doctoral journey.

Moving on to the Micro- and Nanosystems (MNS) Group at ETHZ - the people I have spent every day with the past 5 years:

I am profoundly grateful to Dr. Cosmin Ioan Roman, a senior scientist within our group, whose tireless support and guidance have been instrumental in my academic journey. Cosmin's expertise and dedication have extended beyond mere assistance; he has been a beacon of knowledge, patiently guiding me through data analysis, unraveling the complexities of different simulations, and providing clarity in intrinsically cloudy physics concepts. Moreover, his mentorship has elevated my scientific writing to new heights, by his own hard work and commitment coupled with the occasional laugh about some rare but still occurring exceptionally poorly written sentences. More than a mentor, Cosmin has been a pillar of encouragement during moments of self-doubt, instilling confidence in my abilities with his calm demeanor and trust in me. Without his support, I would not have achieved the milestones I have reached today regarding scientific milestones and publications. Additionally, I am deeply thankful to Cosmin for his compassion and understanding during times of frustration. His ability to calm me down and offer reassurance, even when tears were shed, has been a source of immense comfort and strength. Beyond the realm of academia, I am grateful to Cosmin for the joy and laughter he has brought into our interactions over the years. His infectious humor and warmth have made our collaboration not only productive but also immensely enjoyable. His impact on my academic and personal growth have been immeasurable.

I am also thankful for Dr. Miroslav Haluska, a senior scientist in our group for his engagement of the safety concepts in our group. As the chemical laboratory responsible, I was able to work with him on various different topics. Miro's positive attitude and commitment to help has made him a cherished colleague. His willingness to lend a helping hand and his consistent availability regarding all things safety in the laboratory are a testament to his dedication and reliability. His contributions to the safety of our laboratory have been at the of great importance for the laboratory work within our group.

Furthermore, I would like to thank Manuela Kägi, our secretary who has been at the of our group handling all of our administrative tasks.

I would further like to extend my heartfelt gratitude to Morten Vollmann for his support and guidance throughout my doctoral journey. Morten's contributions have been instrumental in shaping my research, and I am truly grateful for his constant presence. In particular, Morten played an important role in the early stages of my project by helping me in developing fast and efficient Python code for controlling my experimental setup and conducting data analysis. Furthermore, Morten generously shared his knowledge and vast knowledge in utilizing Blender for 3D animation, which significantly improved the quality of the visual representation of my research findings. Moreover, I am deeply appreciative of Morten's support during our discussions about my project. His insightful questions and thoughtful feedback not only helped me gain a clearer perspective on my research but also facilitated effective communication of my results. Morten's input has been at the core in shaping the direction of my research and improving the clarity of my presentations. Additionally, as fellow PhD students starting our journey around the same time, Morten provided emotional support during challenging times. His empathy, understanding, humor, and encouragement have been and still are a source of strength, without which I am unsure I would have come to the point of finishing my PhD. I am grateful for his companionship and friendship throughout our doctoral studies. His wisdom (despite his young age), tenaciousness, and capability of always being positive are a constant support and have been instrumental in my academic and personal growth, and I am truly grateful for his presence in my life.

Specifically, there are some more coworkers I would like to highlight from the MNS group:

Katrina Klösel has been a steadfast companion throughout the trials and tribulations of our PhD journeys. Through both joyful and challenging times, Katrina has been a source of support, and I am grateful for her friendship and resilience. Her ability to find humor in any situation and her loyalty have been constants in my life.

Johannes Weichart has brought immense joy to our workplace with his quirky personality and profound engineering knowledge. His knowledge in microfabrication has been insightful, and his presence has enriched any situation. Beyond his technical prowess, Johannes's positive attitude and willingness to lend a helping hand have made him a cherished colleague.

Ian Mihailovic has been a joyful person to be around in the midst of our academic pursuits, providing not only emotional support but also plenty of laughter with his fun jokes and memes. His camaraderie has made my PhD all the more enjoyable. Ian's ability to lift spirits and see the positive has been a source of happiness in sad situations.

Cristina Gentili has been a constant source of motivation and support, encouraging me to join gym classes even when I've felt reluctant which has had an immensely positive effect on my life. Her presence in the office provides a comforting space where I can freely share both joys and frustrations, knowing she will always lend a supportive ear.

Seoho Jung has been an cherished coworker, offering insightful perspectives and contributing to our team dynamic. Our game nights have been a source of camaraderie and fun, strengthening our bond beyond the confines of our work.

Further, I would also like to highlight coworkers such as Florin Püntener, Sira Bielefeldt, Ines Kraiem, Matthias Dupuch, and Kishan Thodkar Raveendranathrao for the great conversations and times over the years.

I would also like to thank my students Rafael Sonderegger, Loris Togninalli, Christian Thiago Peterhans, Liza Bachmann, Lavinia Recchioni, and Elias Haller who have contributed to many different interesting discussions and results which have enriched the research. Their creativity, attention to detail, and willingness to tackle challenges have inspired me throughout my PhD.

Further, I would like to highlight some of the relationships that have formed in research groups outside of MNS group throughout my PhD:

I am deeply grateful to Suwannee Ganguillet of the Laboratory of Mucosal Immunology for her friendship and support throughout the years. Not only has she been answering many of my *where is this, where is that*-questions in the biological laboratory, but she has also provided well-welcomed company in the later hours in the laboratory when processes go longer than expected. We have been through cackles and tears, celebrated successes and shared sorrows and I would not have been able to enjoy my time in the biological laboratory as much as I have with your friendship. Furthermore, I have to mention the gentle giant Bigfoot who has also provided friendship and support in hard times. I would also like to again thank the whole of the Laboratory of Mucosal Immunology group has been extremely welcoming especially Giorgia, Alice, Louise, Tom, Verena, Swapan, Selma, Niina, and Ronja.

Furthermore, I am grateful to Arnaud Mieville and Konstantin Wolf of the Applied Mechanobiology Laboratory for their assistance in cell culturing and navigating the cell culture laboratory. Their willingness to share knowledge and support has been inspiring and created a safe and collaborative environment. Especially thank you Konstantin, for sharing your cells with me and providing insight into what has worked for you in the laboratory regarding THP-1 cells.

I would also like to thank colleagues at the Product Development Group (PDZ), especially Konstantinos Magkoutas, Dominik Schulte, and Marcel Gort. They have helped with their knowledge and access to 3D printing setups, tensile test measurement setups, and also pulsatile flow setups.

I extend my heartfelt appreciation to the Euler Cluster team at ETH Zurich, particularly Sam, for his efficient and amazing assistance with Slurm and LSF. Their expertise and dedication have been crucial to the success of my simulations. Truly, an environment with *no question is a stupid question* was upheld which helped me to feel comfortable enough to ask questions. Thereby, throughout the years, I managed to develop quite some knowledge and experience using both Slurm and LSF for my simulations. The Euler Cluster's prompt responses and effective troubleshooting has significantly contributed to my simulation results, and I am grateful for their assistance.

I want to thank Donat Scheiwiller, a part of the technical team of FIRST CLA. He has many times helped me out with debugging in the clean room. Donat's knowledge and dedication have been instrumental in overcoming technical challenges. His tips and tricks in the clean room have helped me both save time and also find solutions to problems I have faced.

I want to thank Ute Drechsler and the whole team at BRNC. Their willingness to help and provide guidance have ensured that I have been able to reach the microfabrication goals necessary for my thesis.

I would also like to thank Patrick Lehmann, the leader of the First Aid Team at ETHZ, which I had the privilege of being a part of. I have learned an immense amount of first aid knowledge from the many classes he has organized and also from the situations on duty I have had to use them.

List of Symbols

Table 1: List of symbols, units and a description.

Symbol	Unit	Description
α	-	Womersley number
α_p	1/K	Coefficient of thermal expansion
A	m^2	Surface area
A	Pa/m	Sinusoidal function amplitude
b	m	Channel width
c	m/s	Pulse wave velocity of blood
$c_{p,b}$	J/kgK	Specific heat capacity
C_1	-	Variable
d	m	Arc length
δ	-	Derivative
D	m	Diameter
ϵ	-	Aspect ratio of microfluidic channel
f	Hz	Frequency
$fRe_{\sqrt{\mathrm{A}}}$	-	Friction-factor Reynolds number
F	m^3/s	Flow rate
γ	-	Shape parameter
G	min	Bacterial doubling time
h	m	Channel height
h	W/m^2K	Convective heat transfer coefficient
k	1/s	Slope of exponential fit
k	W/mK	Thermal conductivity
l	m	Length
L_c	-	Ratio of standard deviations
L_{ch}	m	Characteristic length
L_h	m	Hydrodynamic entrance length
L_{th}	m	Thermodynamic entrance length
μ	Pa s	Dynamic viscosity
μ_{OD}	1/s	OD rate of change
μ_Q	1/s	Heat rate of change
∇	-	Nabla operator
N_t	bacteria/mL	Optical density
Nu	-	Nusselt number

Continuation of Table 1		
OD_{ratio}	$\frac{bac/mL}{OD}$	Conversion from OD to bac/mL
π	-	3.1415
P	Pa	Pressure
P_{int}	Pa	Interpolated pressure profile from literature
Pr	-	Prandtl number
ϕ	°	Arc angle
q	W/m^2	Heat flux
$\bar{q}$	W/m^2	Averaged heat flux
$q_{1/or/2}$	W/m^2	Heat flux in channel 1 or 2
q_m	W/m^2	Heat flux for methyl paraben experiments
q_r	W/m^2	Radiative heat flux
Q	W/m^3	General thermal power density
Q_{met}	W/m^3	Metabolic thermal power density of tissue
Q_{ted}	W/m^3	Thermal power density of thermoelastic damping effects
$\dot{Q}$	W	Thermal power/heat flow
$\dot{Q}_{bacterium}$	W	Single bacterium thermal power/heat flow
$\dot{Q}_m$	W	Thermal power/heat flow of microfluidic channel
$\dot{Q}_0$	W	Thermal power/heat flow at the y-axis intercept (t=0)
$\dot{Q}^{\mu}$	W	Thermal power/heat flow measured by the microcal.
r	m	Radius
R	K/W	Lumped element model thermal resistance
R	m	Cylinder radius
Re	-	Reynolds number
ρ	kg/m^3	Density
S	Pa	Stress tensor
σ	W/m^2	Standard deviation
σ_q	W/m^2	Standard deviation of heat flux
t	s	Time
T	K	Temperature
T_b	K	Blood temperature
T_{bulk}	K	Bulk geometry temperature
T_{local}	K	Local temperature
T_{∞}	K	Liquid bulk temperature
τ	Pa	Viscous stress tensor
$\tau_{bacteria}$	1/s	Time constant
u	m/s	Pulsatile element of velocity
u	m	Radial geometrical component

Continuation of Table 1		
v	m/s	Velocity
$\mathbf{v}=(u,w,v)^T$	m/s	Velocity (vector axisymmetric coordinates)
V	m^3	Volume
v_{avg}	m/s	Average velocity
v_z	m/s	Velocity in z direction
v	m	Axial geometrical component
w	m	Azimuthal geometrical component
w_b	1/s	Tissue blood perfusion
ω	1/s	Angular frequency
X_{mp}	%	Heat transfer efficiency
z	m	Distance travelled from inlet

List of Abbreviations

Table 1: List of abbreviations and a description.

Abbreviation	**Description**
BSL2	Biosafety level 2
CD80	Cluster of differentiation 80
CD86	Cluster of differentiation 86
CRMD	Cardiac rhythm management devices
CRP	C-reactive protein
CT	Computerized tomography
DI	Deionized water
DWL	Direct write laser
E. coli	*Escherichia Coli*
EDTA	Ethylenediaminetetraacetic acid
E. faecalis	*Enterococcus faecalis*
ePTFE	Expanded Polytetrafluoroethylene
FACS	Fluorescence Activated Cell Sorting
HLA ABC	Human Leukocyte Antigen, ABC isotype
HLA DR	Human Leukocyte Antigen, DR isotype
ID	Inner diameter
IFN-γ	Interferon gamma
LB	Lysogeny broth
LEM	Lumped Element Model
MG1655	*E. coli strain*
MP	Methyl paraben
MRI	Magnetic resonance imaging
NaOH	Sodium hydroxide
OD	Optical density
PBS	Phosphor buffered saline
PDMS	Polydimethylsiloxane
PET	Positron emission tomorgraphy
PFA	Perfluoroalkoxy alkane
PMA	Phorbol myristate acetate
PMMA	Polymethyl methacrylate
PTFE	Polytetrafluoroethylene
RPMI	Roswell Park Memorial Institute
SD	Standard deviation
TEG	Thermoelectric generator

Continuation of Table 1	
THP-1	Tohoku Hospital Pediatrics-1

Contents

List of Figures

List of Tables

1 Introduction

In this thesis we investigate the application of heat flux sensors for the early detection of infection on implant surfaces. The specific model system investigated is the vascular graft with a growing infection on the tissue side as illustrated in Figure 1.1 [1].

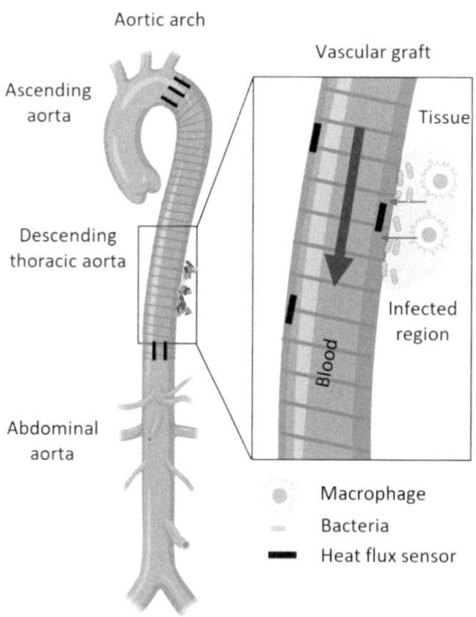

Figure 1.1: Model system of a vascular graft in the descending aorta. The close up shows the infectious growth and the position at which the heat flux sensors are positioned for the detection of infectious growth onset. Open Access figure adapted from [1] published under open access under CC BY license (Creative Commons Attribution 4.0 International license). Figure created with Biorender.

Implant infections on vascular grafts are relatively rare; however, with the increasing number of patients receiving implants, complications such as infections also rise [2]. Current infection detection methods are often late-stage and costly, prompting the need for earlier detection of bacterial growth or infections on implant surfaces. Early detection is crucial, as the formation of a biofilm reduces significantly the effectiveness of antibiotic treatments.

While implant infections alone occur in only 4% of aortic vascular grafts [3], the escalating number of patients with cardiovascular complications leads to a greater overall incidence of implant infections [2]. A rising trend of infection rates on cardiac implants has been observed [2]. Mortality increases drastically upon vascular graft implant infection, ranging from 17% to 40% of patients [4]. Given the susceptibility of implant surfaces to bacterial infections [5], early detection of infection is thus of increasing importance.

Different sensing approaches using chemical and biochemical sensors have been investigated for the detection of infection. However, due to their limitation of functionality due to drift and aeging *in vivo* [6], physical sensors show higher potential for stable long-term sensing. Temperature sensing has been investigated as a potential infection indicator, as already indicated in the first century CE in *de Medicina* by Celcus [7]. However, a lesser explored avenue is heat flux sensing for infection detection. This physical sensing mode is less affected by the foreign body response and is also less susceptible to environmental temperature variations.

The aim of this thesis is to assess the viability of utilizing heat flux sensing as an indicator for the initial stages of infection through a combination of experimental investigation and modeling techniques.

1.1 Objective and outline of this thesis

The objectives and results of this thesis can be summarized as follows:

- State of the art literature review on markers for infection.
    - It will be concluded that the heat flux is a viable and yet to be explored marker for infection on implant surfaces.

- Identification of the main sources of heat flow at an infected implant surface.
    - The main individual sources of heat flow and measurement methods are summarized:
        * Bacteria
        * Immune cells
        * Blood perfusion

- Development of the fabrication steps of an *in vitro* microfluidic calorimetric chip allowing for the experimental *in vitro* determination of the limit of detection of bacterial growth.

- Development of an experimental protocol for the thermal calibration of the microfluidic chip using chemical exothermic reactions.
    - Comparison of microfluidic chip data with microcalorimetry data.
    - Determination of the heat transfer efficiency of the microfluidic chip.
    - Demonstration of the versatility of the microfluidic chip with a double channel experimental setup mimicking the thermal environment of blood flow in the aorta.

- Development of an experimental protocol for the measurement of heat flux originating from the bacterial growth of *E. coli* and the determination of the single *E. coli* bacterial thermal power.
    - Determination of the thermal resolution of the *in vitro* microfluidic chip in relation to a population of bacteria.
    - Determination of the limit of detection of the bacterial growth in the chip.
    - Demonstration of a double channel *in vitro* microfluidic chip mimicking the specific aortic vascular graft thermal environment.

  - Demonstration of the detection of bacterial growth in the *in vitro* microfluidic chip system in a thermally un-stabilized environment to further mimic the thermally fluctuating environment expected around an implant *in vivo.*

- Investigation of immune cells as a source of heat flow at infectious sites.
  - Creation of protocols for the stimulation pathways of immune cells for gram-positive and gram-negative bacteria.
  - Microcalorimetric measurements and FACS experiments of stimulated immune cells.
    * Determination of measured heat flow in response to different stimulants: cytokines and bacteria (*Escherichia Coli*/*Enterococcus Faecalis*).
    * Determination of the cell types and the stimulation response of the immune cells using FACS.

- Modelling and simulation of the heat transfer environment of both an infection growing on a vascular graft and of the *in vitro* microfluidic chip.
  - Investigation of the influence of pulsatile flow on the heat transfer coefficient in the aorta.
  - Determination of the viability of using heat flux sensors for the early detection of infection.
  - Estimation of the density of a sensor array of heat flux sensors on the surface of a vascular graft for full coverage infection detection.
  - Experimental verification of the simulation results.

- Conclusion towards the *in vitro* feasibility for using heat flux sensors as a marker for the early detection of infection on implant surfaces .

This thesis is structured as follows:

Chapter 2 State of the Art: The state of the art of infection detection pathways in a clinical environment is presented. The origin of infectious growth development on implant surfaces is described in relation to how infections are currently detected in patients with implants, specifically on vascular graft implants. Further, different approaches to infectious sensing systems and inflammatory markers in research are explored. The feasibility of heat flux as an infectious marker is investigated. Specific heat flow sources considered at the infectious site are presented. A summary of literature on microcalorimetric systems for the thermal measurements of different cells in relation to infectious heat flow sources is reviewed. The implementation of heat flux measurements on microfluidic chips are discussed as platforms for the characterization of thermal power of biological samples. Further, a literature review of the fluid dynamic considerations regarding the vascular graft model system is presented and how blood pulsatile flow influences the convective thermal transfer at the implant surface is discussed. In this chapter, the basis for the further investigations throughout the different chapters is set.

Chapter 3 Theory and Modelling: The theoretical aspects of the hydrodynamic and thermodynamic considerations of a model system of the aortic vascular graft implant are presented. The theoretical background of the influence of pulsatile flow on the heat transfer coefficient in laminar, low and high Reynolds number flow environments is presented. Basic thermodynamic concepts relevant to the model system of the aortic vascular graft and the investigated *in vitro* microfluidic chip are introduced. Further, the concept of Lumped Element Models is introduced and applied to the different variations of the microfluidic chips. The theoretical background for the finite element modelling of fluid dynamic and thermodynamic simulations is described in detail.

Chapter 4 Fabrication: The fabrication of the microfluidic chips is presented with different experimental designs for the chemical calibration, bacterial experiments and immune cells. The full bottom-up approach using soft lithography for the fabrication of PDMS channels, the assembly of the microfluidic chip with the integrated sensors, and the experimental protocols are described. Further, the biological protocols for bacterial cell culture and immune cell handling/stimulation are listed and described in regards to general handling and also in relation to the microfluidic chip.

Chapter 5 Experimental Results: All of the experimental data for the microfluidic chip experiments are presented. Starting with the thermal calibration using the exothermic first order reaction of the base catalyzed hydrolysis of methyl paraben and the measurement of the exponential growth curve of *E. coli*. The differential sensing system with common mode rejection of the thermal background signals is discussed and the extraction of relevant heat flux signal is described in detail. Further, the determination of the thermal power of an *E. coli* bacteria is shown through the measurement of heat flux and the parallel measurement of the bacterial population through optical density measurements. Extension of the microfluidic chip to a double channel chip to further emulating the thermal environment of blood flow at the aortic vascular graft is shown. In addition to this, experiments in thermally not stabilized environments with sensors of worse resolution and smaller size to further emulate more realistic thermal environment expected *in vivo* are also shown. Further, investigations of the thermal power originating from immune cells stimulated by cytokines and bacteria is determined using FACS analysis and microcalorimetric measurements.

Chapter 6 Simulations: Simulations of the thermodynamic and hydrodynamic environment surrounding the vascular graft implant are presented. The influence of pulsatile flow in aortic vascular grafts on the thermal environment is quantified by determining the heat transfer coefficient. Parametric sweeps of graft material, implant size, flow velocity, frequency, and amplitude are investigated. The thermal power and mass of an infection at which detection through a heat flux sensor is possible is determined.

Chapter 7 Conclusion and Outlook: The feasibility of applying heat flux sensors as a means of early detection of infection on an aortic vascular graft is explored. Agreement between the experiments and the simulated heat flux resolution of the system is demonstrated. Further, the detection limit (in simulation and experiments) is shown to be at a small fraction of the postulated thermal power of a bacterial biofilm. These results support the hypothesis of successful early detection of infection using heat flux sensors on vascular graft implants.

2 State of the Art

A color version of this chapter is available at https://doi.org/10.3929/ethz-b-000703739.

2.1 Infections on Implant Surfaces

The biological mechanism determining whether or not an implant is infected or not is often referred to as *the race to the surface* [8,9]. The implant surface compromises the defensive mechanism of the body locally, which facilitates local bacterial adhesion and growth [5]. It has previously been shown that fewer bacteria are necessary to infect a surgical site in the presence of an implant surface in comparison to without an implant [9]. As implant surfaces are thereby especially prone to infections and increasing amounts of implants are used in medicine, different ways for the early detection and treatment of implant infection is becoming of increasing importance in research.

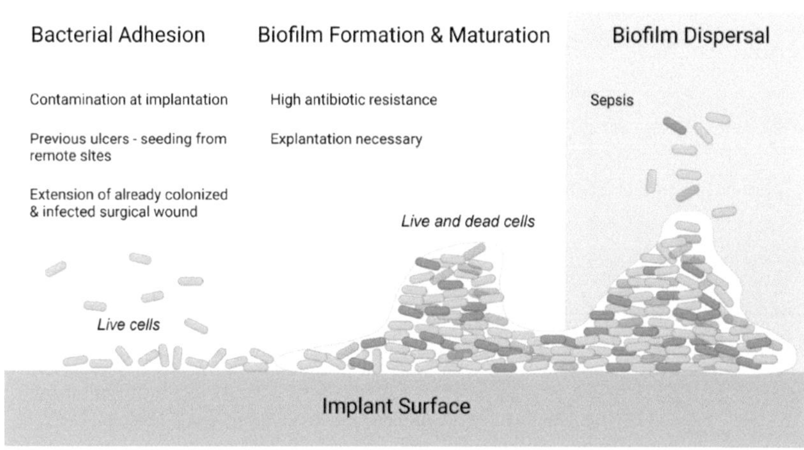

Figure 2.1: Infectious growth at the implant surface. Figure drawn based on literature [5] and created with BioRender.com.

The development of an infection on an implant surface can be divided in 3 different stages; bacterial adhesion, biofilm formation and maturation, and biofilm dispersal [5] as shown in Figure 2.1. The initial adhesion on the implant surface is governed by physiochemical interactions between the bacteria and the implant surface [10]. The source of the bacteria could be during the surgery or seeded remotely from another infected site in the body [8]. Upon adhesion, the bacteria proceeds to form a biofilm matrix made of extracellular polymeric substances with which it surrounds itself [11]. This polymeric matrix protects the bacteria from conventional treatment, allowing for the biofilm to become more resistant to antibiotics [12]. Upon biofilm formation, the biofilm will subsequently mature and spread itself on the implant surface. At this point the biofilm develops high antibiotic resistance and an explantation of the implant is necessary. If left untreated, the biofilm grows further and could rupture and cause biofilm dispersal into the host tissue and blood. A ruptured biofilm could lead to systemic infection/sepsis if the bacteria reaches the blood flow [9, 13]. The most common bacteria for orthopedic implants and for prosthetic vascular grafts infections are Staphylococci [10, 14]. Specifically on aortic vascular grafts, the most common bacteria found are Staphylococcus aureus, Escherichia coli, Enterococcus spp., and Pseudomonas aeruginosa which is a mix of gram positive and gram negative bacteria [15].

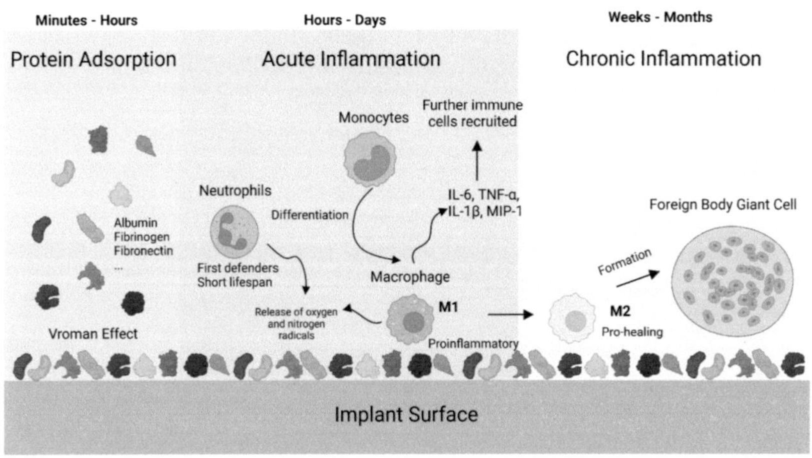

Figure 2.2: Inflammatory response of the body to an implant. Figure drawn based on literature [16] and created with BioRender.com.

The *foreign body response* consists of multiple steps, which unfold from days to years as shown in Figure 2.2. The duration and severity of this response is dependent on the implant location, the surface chemistry, and the functionality of the implant [17]. In all cases, the process starts with an initial exposure to blood which initiates the so-called Vroman effect. This is a molecular surface adsorption cascade which triggers protein adsorption from blood where smaller molecules are replaced by larger molecules over time. After this initial protein layer, different immune cells gather around the implant, and following a response over weeks or even months, the *foreign body response* ends with either rejection of the implant, healing impairment, tissue damage, collagenous encapsulation, infection, or chronic inflammation.

The introduction and also presence of a foreign body/implant causes injury in the body. As already introduced, the interaction between the implant and blood follows a molecular surface adsorption scheme called the Vroman effect [18] as illustrated in Figure 2.2. Leo Vroman introduced the concept of time-dependent adsorption of proteins present in the blood replacing each other over time at liquid/solid interfaces [19]. According to this time and surface-material dependent adsorption, the smallest proteins adsorb first, sequentially being replaced by larger proteins over time. This diffusion dependent process follows the following order of adsorption: albumin, immunoglobulin G (IgG), fibrinogen, and lastly high-molecular-weight kininogen (HMWK) [19]. Due to the dependency of the protein adsorption on the surface material of the implant, the foreign body response can be influenced by material choice within the first minutes and hours within the body [20]. The matrix created of adsorbed proteins dictate the cellular interaction with the implant and not directly the implant material itself [21]. This also has an impact on the quality of wound healing around the implant as the protein matrix creates a basis for healing and tissue regeneration. Understanding and leveraging the protein adsorption on the implant surface therefore is a powerful tool to modify the *foreign body response.*

In a next step, inflammatory cells such as neutrophils gather around the protein matrix formed on the implant surface [22]. The amount of neutrophils that gather around the implant surface dictate the amount of monocytes and macrophages at the implant site [23]. The neutrophils are the body's first line of defense [23] and are an indicator of acute inflammation, which lasts from hours to days after implantation as shown in Figure 2.2. A persisting acute inflammation longer than three weeks is an indication of an infected implant [22]. Neutrophils try to phagocytise any type of foreign material in the body, as their main task is to clean the wound site [24]. If the foreign material in question is in a similar size range as the neutrophil, it is phagocytised and degraded by the cell. However, most implants are orders of magnitudes larger than the cell, which can cause frustration of the phagocytosis [25] due to size disparity. This causes the production of peroxides which increase the biodegradation of the implant material [26], which especially becomes an issue with polymeric materials. As a further step, the neutrophils release enzymes which influence the biostability of the implant material due to their corrosive properties [27].

When other inflammatory cells than neutrophils are recruited to the implant surface, the inflammation transitions from acute to chronic inflammation. In contrast to acute inflammation, chronic inflammation is not indicated by strong symptoms like acute inflammation and last over a longer time period than acute inflammation. The other inflammatory cells present during chronic inflammation include monocytes which differentiate to macrophages [25] (as illustrated in Figure 2.2). Macrophages are able to produce biological components such as reactive oxygen species (ROX), growth promoting factors, coagulation factor, and cytokines [24]. The release of ROX further contributes to the corrosion of the implant material. Furthermore, the release of some of these biological components such as cytokines, which include monokins, interleukins, and chemokines, trigger further immuno system defense mechanisms [28]. Specifically, the presence of transforming growth factor β (TGFβ) triggers the transition of macrophages to myofibroblasts [28]. However, chronic inflammation does not always occur and its severity is dependent on the implant material (chemical and physical properties) and location of the implant [29].

2.2 Current Implant Infection Detection

Infections on implants manifest themselves in patients through different symptoms depending on the time passed between surgery and the onset of infection as illustrated in Figure 2.3. Early infection, defined as within 3 months of surgery, often manifests itself with concrete symptoms such as fever, pain, swelling and a pulsating mass [5]. Predominantly, early infection is due to contamination during surgery [10] and late infection can initially be asymptomatic and is either caused by less virulent organisms or seeding form remote infections in the body [5]. Currently, the clinical diagnosis is by CT imaging, ultrasound, magnetic resonance (MRI), positron emission tomography (PET) or indium white blood cell (WBC) scan [14]. Computerized tomography (CT) imaging shows periprosthetic fluid collection, however the image quality around the prosthesis is often limited [4, 14]. MRI has an increased resolution, however it carries risk factors for certain patients with renal insufficiency [4]. It is often used when the CT scan does not give conclusive answers, however it is more expensive [14]. PET scan is also used to identify vascular graft infections, however it is even more expensive than other imaging methods. Indium WBC scan is used in combination with other imaging results and has the lowest sensitivity. It is used when CT and MRI are inconclusive, in this case obtaining the results take more than 24 hours [14].

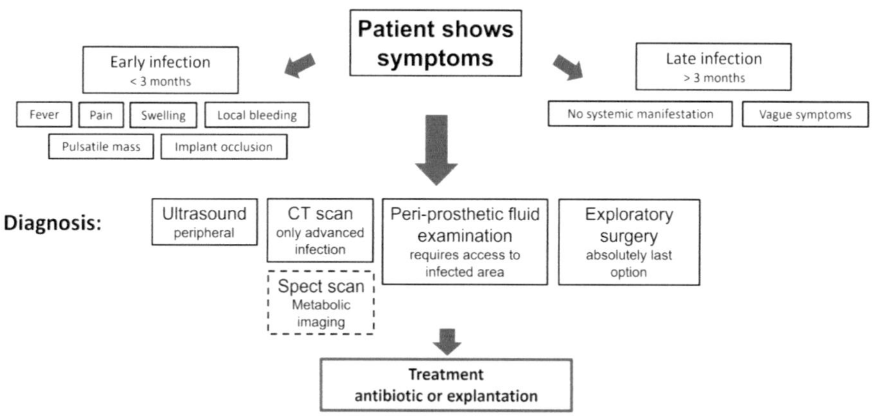

Figure 2.3: Infection detection of patients with implants. Sketch based on literature [5, 14].

The common issue with these imaging methods for the detection of infection is the restricted access to localized information and inconclusiveness. Furthermore, an indication of the onset of infection prior to biofilm formation is not possible, as imaging methods are only applied after suspected vascular graft infections [4] which manifests itself in late stage symptoms. Thereby, an implanted sensor for localized sensing of an inflammatory marker could aid in the early diagnosis and thus allow for more efficient treatment of infection on implant surfaces [6].

2.3 Inflammatory Markers

Various different biomarkers have been investigated for the detection of infection such as C-reactive protein (CRP) [30–32], lactate [33–35], glucose [33–35], pH [36–38] and temperature [39, 40]. CRP is integral for the activation of macrophages and it is triggered to be produced by the immune system upon inflammation. Studies have shown that CRP concentration in blood also is an early marker for prosthetic hip and knee infection [30]. It has been shown to be an indicator of inflammation as the local concentration of CRP changes with inflammation [35, 41]. Both changes in the values for glucose and lactate have been related to the inflammatory response in wounds, as both are indications of a change in the metabolism in the wound [35]. Changes in pH values are related to the potential development of infection on a wound site [38]. Furthermore, temperature increases in wounds have also been related to wound infection [39], in some studies the rise in temperature has even been shown to be predictive for foot ulcerations [40]. These biomarkers are often investigated either already during diagnosis or in research for the potential use in point of care diagnostics.

As listed in Table 2.1 lactate and glucose have been investigated in an *in vivo* mouse model showing changes in both values with a local bacterial infection [34]. This initial study shows promising potential of chemical sensors for the detection of localized infection *in vivo*. CRP has been studied *in vitro* as a surgical catheter for the live time surgical infection sensing [31]. Multiple wearable pH sensing systems have been investigated for wound monitoring [38]. One study also shows an orthopedic implant with an integrated pH sensor [42]. These mentioned sensor systems have been investigated on *in vitro* systems. Implantable temperature sensors have been investigated both *in vivo* and *in vitro* in systems such as medical sutures [43], wearables [38], and in orthopedic implants [44]. The foreign body response limits the performance of chemical sensors by encapsulating the sensor [6, 45, 46]. Thus, chemical and biochemical sensors have the limitation of drift, as the sensing of the target component is dependent on diffusion of the molecule/biomarker to the sensor surface. Furthermore, ageing and poisoning (chemical changes) of the sensor functional materials also alter the performance of chemical sensors. Physical sensors such as heat flux sensors or temperature sensors therefore have higher potential for stable long-term sensing.

In general, the specificity of chemical sensors is higher than physical sensors due to the possibility of measuring targeted biomarkers such as antigens or proteins. This is leveraged in cancer research for example for ovarian, breast and gastrointestinal cancer [49]. The specific chemical marker indicating an inflammation in the body is CRP concentration in blood [32]. CRP concentration increases with inflammation [32]. Changes in glucose and lactate are related to a general localized inflammation, however, they are not specific indicators for infection or a specific pathogen [34]. Similarly, changes in pH is related to abscess formation (decreasing pH in the environment), wound healing [50] and also biofilm formation [42, 51]. Nevertheless, changes in pH are also not related to a specific infection. Temperature changes occur upon inflammation, which includes acute inflammation (an initial step of wound healing within the first couple of days [52]) and pathogenic infections (first symptoms between the first month and 2 years [10]). Thereby, even though less specific than chemical markers it has the potential of indicating infections on implant surfaces while retaining long term sensing properties.

Table 2.1: Table with a literature review over different inflammatory markers researched in literature. T is temperature, O_2 is the oxygen, lac. is lactate, and gluc. is glucose.

Literature on Inflammatory Markers in Research							
Marker	Measurand	System	Principle	*in vivo*/*in vitro*	Author	Journal	Year
Temp.	Local infection in orthopedics	Thermistor embedded in screw	Increase in T with infection	Test of sensing capabilities	Karipott et al. [44]	IEEE Sensors Jounral	2018
Temp.	Wound monitoring	Medical suture	General diagnostic purposes	*in vivo* (animal)	Kim et al. [43]	Small	2012
Temp.	Skin thermal characterization	wearable and flexible	General diagnostic purposes	*in vivo*	Zhang et al. [47]	Nature Materials	2013
Temp. & pH	Food ulcers/diabetes	Wearable	Increase in T and pH	Test of sensing capabilities	Salvo et al. [38]	Int. J. Nanomed.	2017
pH	Orthopedic implant	On implant surface	Increase in pH	*in vitro* tissue sample	Uzair et al. [42]	ACS Sensors	2019
pH	Wound	wearable	Increase in pH	Test of sensing capabilities	Yun et al. [48]	IEEE Sensors	2017
pH	Wound	wearable and flexible	Increase in pH with infection	Test of sensing capabilities	Santos et al. [37]	ACS appl. Mater-Interfaces	2014
O_2, lac. & gluc.	Bacterial infection	Implantable	Increase in pH	*in vivo* mouse model with S. Aureus	Ring et al. [34]	Biomed. Engineering	2017
CRP	Bacterial inflammation in blood	Surgical catheter	Increase in CRP	*in vitro*	Ji et al. [31]	Advanced Science	2018

2.4 Heat Flux as an Infectious Marker

In the 1st century CE, a philosopher named Celsus identified the four signs of inflammation in his book De Medicina: "*calor (heat), dolor (pain), rubor (redness) and tumor (swelling)*" [7]. A localized increase in temperature (and in severe cases a global increase in temperature) has thereby for centuries been known as a marker for inflammation/infection. Looking at an infectious site on the micro-scale, the sources of heat flow can be determined as increased metabolic activity of the immune cell due to phagocytosis, bacterial cell replication tissue damage, and increased blood flow to the inflamed region as illustrated in Figure 2.4.

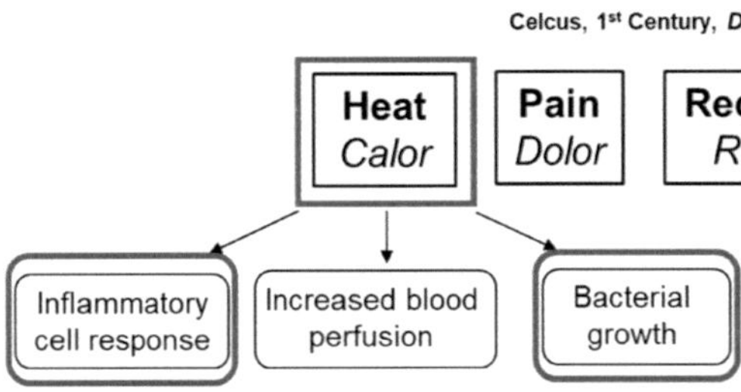

Figure 2.4: Symptoms of infectious growth as described by De Medicina [7] and the postulated individual heat flow sources. In this thesis, the heat flow by bacterial growth and by the inflammatory response of immune cells is in focus.

Most biological processes are exothermic, meaning heat is produced [53, 54] and it can even be broken down into sets of metabolic pathways [55]. These biological processes heat up the cells and the immediate environment around the cells. In the body, the metabolic heat flow originating form these cells can be measured as body temperature on a macroscale and is closely regulated. Specifically, at an infection site heat flow originates from activated immune cells (neutrophils, monocytes and macrophages) and bacterial cell replication. The bacteria innately produce metabolic thermal power while growing and it has even been shown that the growth curve of different bacterial strains can be distinguished by their heat flow curves [56]. On the other hand, the immune cells have many different catabolic processes (energy releasing processes as illustrated in Figure 2.5) which increase, upon their exposure to bacteria or other pathogens [57]. The immune cells identify pathogens through their pathogen recognition receptors and phagocytose these pathogens [58, 59]. They also release oxygen and nitrogen radicals [59] in an attempt to suppress the growth and spread of the bacteria in addition to phagocytosing them [59].

Macrophages are derived from differentiated monocytes, and this can be controlled through stimulation. It can be classified in two different types, either M1 macrophages or M2 macrophages. M1 macrophages respond to pathogen-associated molecular patterns (PAMPs) and produce pro-inflammatory cytokines by PRRs [60]. They are the first line of protection against pathogens such as viruses and bacteria. M2 are activated by cytokines and are active in the process of wound healing. The immune response of macrophages is studied in literature in microcalorimeters by adding pathogens such as a virus [61], coated nanoparticles [62], chemicals like polyhydroxylkanoate (PHA) [63, 64] or lipopolysaccharides (LPS) [65, 66]. THP-1 cells have been shown to be a good system to model the response of monocyte-macrophage differentiation [67] and has been generally accepted as an *in vitro* system for immune modulation [68].

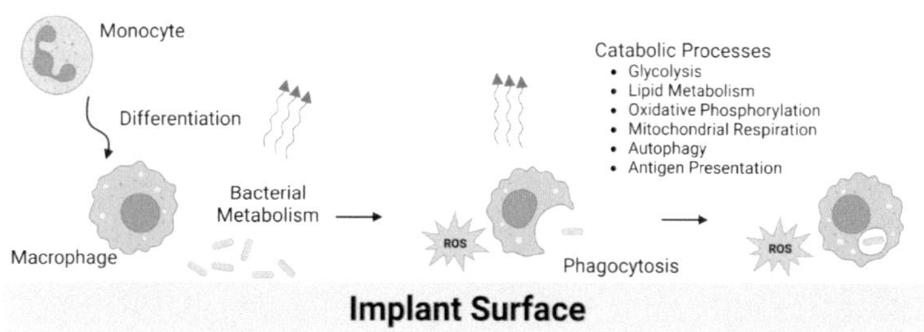

Figure 2.5: A schematic over the catabolic processes for immune cells and bacteria when they interact. Based on information from literature [59]. Sketch created with BioRender.com.

Many orders of magnitude larger than the immune cells and the bacteria, infected wounds and non-infected wounds show a temperature difference of over 3 K [39]. Furthermore, the onset of foot ulceration (inflammation) can be indicated by an increase in temperature up to a week before the foot ulcer develops [40]. The increased metabolism of the immune cells at an infection in addition to the bacterial growth increases the temperature as already discussed by de Medicina centuries ago [7]. Furthermore, in simulations the metabolic heat flow originating from infected regions are given a higher thermal power per cubic meter value than the rest of the body [69–71], indicating a temperature gradient in the simulated tissue model. Thereby, considering localized heat flow increases as a potential biomarker for infection is reasonable.

2.5 Measurement of Metabolic Changes and Heat Flow of Biological Samples

Microcalorimetry is the standard measurement method of the metabolic heat flow of all types of cells. Microcalorimetry is commonly used in microbiology [72, 73], soil science [74], food science [75], and medicine [76]. For example, the measured metabolic heat flow of a hepatocyte (liver cell) is orders of magnitude higher (330 pW $cell^{-1}$) [77] than that of a human blood cell (1 pW $cell^{-1}$ - 5 pW $cell^{-1}$) [78] or an alveolar macrophage (19.4 pW $cell^{-1}$) [79]. The size of these cells also vary between 6200 μm^3 [80], 206 μm^3 [81] and 1328 μm^3 [82] respectively.

However, these microcalorimeters are expensive, large, and bulky. Furthermore, they do not allow for any parallel investigations of the samples in addition to heat flow measurements. To overcome this, different chips with calorimetric properties have been researched and developed, specifically, microcalorimetric chips [83–87] and nanocalorimetric chips [54, 85]. These investigated systems include open-type [88–90] and closed-type chips, flow-through or static measurements, and single measurements or twin measurements [54]. Lerchner et al. investigated multiple variations of calorimetric measurements such as a biofilm on a chip [91, 92] and droplet of bacteria in a flow-through chip microcalorimeter [93]. The current investigations focus on the miniaturization of large microcalorimeters with a focus on accuracy and multiple additional parts such as PID controlled heaters, and thermally insulating layers to ensure thermal stability [94, 95]. These additional parts are large and bulky, and limit further design flexibility of the measurement platform. An overview over these developed chip systems is shown in Table 2.2

Table 2.2: Literature on microcalorimetric chip systems summarizing their main properties.

Microcalorimetric Chip Systems					
Author	Sample size [µL]	Heat flow density [W/L]	Open/ closed	Material	Sample measured
Wonhee Lee et al. [87]	0.0035	1.2	Closed/ flow	Parylene, SU8, PDMS, vacuum	Urea hydrolysis catalyzed by urease
Johannes Lerchner et al. [96]	12	0.004	Open	PMMA, Si_xN_x, SU8	Human sickle-cell erythrocytes
Yuan Jia et al., 2015, [97]	10	0.021	Closed	Polyimide, PDMS	Lysosyme, protein unfolding
Frida Mariana et al. [92]	10	0.005	Closed	PMMA, Si_xN_x, SU8	Biofilm reaction to antibiotics

Changes in the heat flow output of cells indicate changes in metabolic behavior which is a marker for successful drug delivery, phagocytosis, or cell differentiation. These investigations have been done on lab-on-chip systems for immunology [98], diagnostics [99, 100], and pharmacology [101]. Specifically, systems to mimic tumors [102], or the lung [103] or kidney [104], and the drug efficiency against bacteria [105] or of cancer biopsies [106] have been investigated. These approaches allow the investigation in *in vivo* environments on small-scale *in vitro* chip systems. In general, these lab-on-chip systems have been developed to investigate microscale changes based on specific biochemical markers or optical inspections with fluorescence. Processes such as glucose respiration, glucose fermentation and lactose fermentation have been measured in microcalorimeters with bacterial populations. They can be used to detect contaminated samples in food science [75] or even aid in the detection of infected blood samples in medicine [76]. These thermograms reflect the chemical conversion of energy by each individual cell [55], describable by Hess's law [107]. The thermal power originating from growing bacterial cells can be correlated to the different processes with the additional knowledge of the available oxygen and glucose in the media [72].

As current micro- and nanocalorimetric chips, are designed as scaled-down versions of traditional instrumentation microcalorimeters, they prioritize measurement accuracy over design flexibility. This focus hampers their adaptation for measuring metabolic heat flow changes in complex microfluidic systems such as lab-on-chip or organ-on-chip platforms. Despite advancements, there remains a gap in integrating chip properties with measurements of cellular metabolic changes through cellular heat flow output. This prompts the exploration of a new approach to measuring biological thermal power output on microfluidic chips with flexible fabrication steps.

2.6 Fluid Dynamic Aspect

As our model system is a vascular graft, the heat transfer is heavily influenced by the convective heat transport at the interface between the vascular graft and the blood. To determine the feasibility of the application of a heat flux sensor as shown in Figure 1.1, the thermal environment of an infectious growth on a vascular graft must be analyzed. There are two directions that the localized heat flow originating from the infection can flow - either towards the tissue or in the direction of the implant, embedded heat flux sensor and the blood flow. In the aortic system, the blood flow effectively has the property of a heat drain, and the heat transfer in the direction of blood flow can be defined by the heat transfer coefficient.

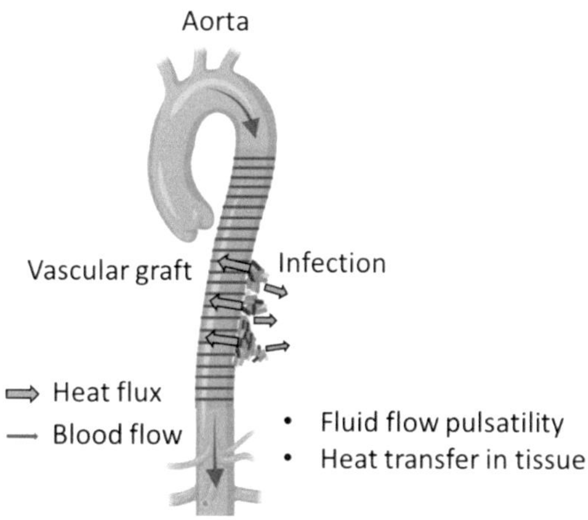

Figure 2.6: Schematic overview over the system investigated through finite-element-modelling (discussed in chapter 6). Sketch created with BioRender.com.

One of the main properties of blood flow is the dynamic pulsatile flow. This type of flow influences the heat transfer environment at the infectious site and also the amount a heat flux sensor would be able to sense in an aortic vascular graft. The influence of laminar pulsatile flow in the different flow regions (developing flow, hydrodynamically fully developed and thermally undeveloped flow, and fully developed flow) on the heat transfer coefficient has previously been studied in literature [108]. It is important to determine and distinguish the different fluid flow regions as they dictate the heat transfer. Numerical investigations by Chattopadhyay et al. [109] and Kim et al. [110] show that there is no significant influence of flow pulsatility in the laminar developing flow region on the heat transfer coefficient. Furthermore, in the fully developed flow region,

Hemida et al. [111] and Yu et al. [112] found no significant influence of pulsatile flow on the heat transfer coefficient for their studied parameters. However, different studies by Habib et al. [113] have shown that the frequency influences whether or not the heat transfer coefficient increases or decreases. They show an increase up to 40% with certain frequencies and a decrease of up to 20% with other frequencies. A smaller 6% decrease in the heat transfer coefficient has been found for the hydrodynamically fully developed and thermally developing region as shown by Hemida et al. [111]. Chattopadhyay et al. [109] found no influence by the pulsatility on the heat transfer coefficient, however, they determined a large influence of the distance from the inlet on the heat transfer coefficient. There is limited literature on the more relevant flow region of turbulent flow (especially low-Reynolds number turbulent flow). However, despite the numerical and experimental difficulties of the studies in these flow environments, Zhao et al. [112] have investigated the influence of pulsatility on the heat transfer coefficient.

These studies are mainly focused on general investigations of pulsatile flow and their influence on the heat transfer coefficient. These investigations are relevant for pipe flow investigations for larger scale industrial applications. Fewer papers focus on biofluid dynamic flow conditions. A paper by Craciunescu et al. [114] details the blood flow conditions and its influence of the surrounding temperature distribution of a dog aorta. They find that the influence of pulsatile flow is minimal on the surrounding. However, due to the smaller aortic radius, this investigation is in the laminar flow condition. Thereby, investigations on the thermodynamic environment of the aortic vascular graft are still to be researched.

2.7 Approach of this Thesis

In this thesis, the feasibility of using heat flux sensors for the early detection of infection on implant surfaces, specifically aortic vascular grafts is investigated. The research question is approached by building an *in vitro* microfluidic chip and isolating the measurement of different determined thermal power sources such as bacteria and immune cells. With this system the thermal resolution (amount of cells, thermal density, heat flux etc.) above which a growing infection can be measured by a heat flux sensor is determined. Further, the microfluidic chip is exposed to multiple variations to determine the feasibility of the microfluidic chip as an *in vitro* system mimicking the thermal environment of the aortic vascular graft. With the experimentally determined thermal resolutions, the sensing possibilities are explored by Finite Element Models in COMSOL to model the *in vivo* environment of the aortic vascular graft. This allows for conclusions on the possibilities of using heat flux sensors for the early detection of infection on a vascular graft surfaces.

3 Theory

In this thesis, the hypothesis of whether or not the early detection of infection on an implant surface is possible through heat flux sensing is investigated. To answer this question, different models, equations, chemical calibration systems, and cell growth systems in different experimental settings are researched. Principally, two different main systems; the aortic vascular graft and an *in vitro* microfluidic model mimicking the infectious bacterial growth on the implant surface are investigated. In both these systems, it is important to understand and characterize the different types of flow and heat transfer as sketched in Figure 3.1. Conduction and convection are present in the system, both of which will be discussed extensively in both systems. To understand the convective heat transfer one also need to understand the fluid flow environment present in the respective systems. The reigning equations in both the fluid, solid and at the interface between them are determined. When the different paths of heat transfer are characterized in the system, one can further estimate the heat transfer efficiency through a lumped element model. Moreover, the theoretical framework for implementing a finite element model of pulsatile aortic blood flow and its impact on convective heat transfer is shown. The different parameters used in this thesis are listed in the Appendix chapter A.1. A color version of this chapter is available at https://doi.org/10.3929/ethz-b-000703739.

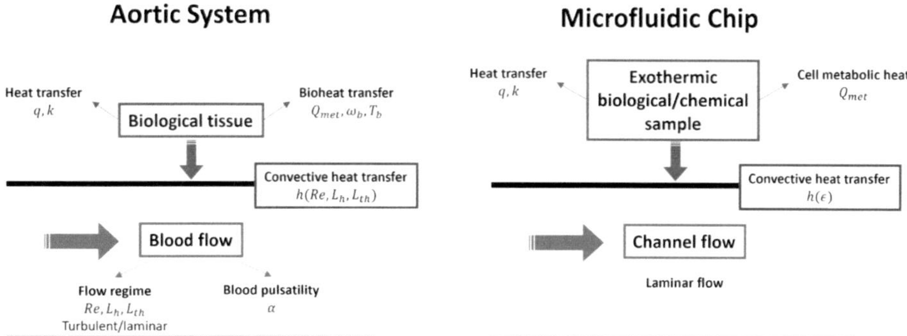

Figure 3.1: Schematic showing the fluid dynamic and thermodynamic aspects considered in both of the systems.

3.1 Fluid Dynamics

3.1.1 Navier-Stokes Equations

The Navier-Stokes equations (continuity and momentum equation) mathematically describe fluid flow in different environments and can be modified for different boundary conditions. The general theory introduced in this section is based on literature [115, 116].

The continuity equation which dictates the mass conservation of incompressible fluids (ρ is constant) is expressed as follows:

$$\nabla \cdot \mathbf{v} = 0 \tag{3.1}$$

The momentum equation can be described through:

$$\frac{\partial \rho}{\partial t}\mathbf{v} + \rho(\mathbf{v} \cdot \nabla)\mathbf{v} = -\nabla P + \mu \nabla^2 \mathbf{v} \tag{3.2}$$

With ρ [kg/m^3] as the fluid density, $\mathbf{v}$ [m/s] as the fluid velocity vector, P [Pa] as the fluid pressure, and μ [Pa s] as the dynamic viscosity.

These equations govern the description of fluid flow and can be modified for different flow conditions.

3.1.1.1 Fluid Flow in Cylinders

The geometrical structure of a descending aorta can be simplified as a cylinder. With a cylinder of a certain radius R, the fluid flow can be described through the axisymmetric coordinates (r, θ, z). The continuity equation 3.1 is thus expressed as [115, 116]:

$$\frac{1}{r}\frac{\partial}{\partial r}ru + \frac{1}{r}\frac{\partial}{\partial \theta}w + \frac{\partial}{\partial z}v = 0 \tag{3.3}$$

With (u, w, v) as the radial, azimuthal and axial components of the velocity vector $\mathbf{v}$ and (r, θ, z) as the axisymmetric coordinates. This can be simplified using the intrinsic symmetry of a cylindrical structure to ($\frac{\partial w}{\partial \theta} = \frac{\partial v}{\partial z} = 0$):

$$\frac{\partial}{\partial r}(ru) = 0 \tag{3.4}$$

The momentum equation 3.2 is expressed for the axial component of **v** (v) as:

$$\rho\frac{\partial}{\partial t}v = -\frac{\partial}{\partial z}P + \mu\frac{1}{r}\frac{\partial}{\partial r}(r\frac{\partial}{\partial r}v) \tag{3.5}$$

With ρ [kg/m^3] as the fluid density, v [m/s] as the axial component of the fluid velocity, P [Pa] as the fluid pressure, μ [Pa s] as the dynamic viscosity, t [s] as the time, and r [m] as the radial component of the axisymmetric coordinates. Adding the knowledge of symmetry, assuming stationary flow, with the conclusion that the pressure gradient is constant and enforcing the boundary condition, one is left with:

$$v(r) = \frac{1}{4\mu}\frac{\partial P}{\partial z}(r^2 - R^2) \tag{3.6}$$

This equation describes the Poseuille flow in a pipe of the axial component of the velocity and shows the well-known parabolic shape of the velocity profile in a pipe. The volume flow rate, F [m^3/s], can be described by the integral:

$$F = 2\pi\int_0^R rv(r)dr \tag{3.7}$$

Which gives:

$$F = -\frac{R^4\pi}{8\mu}\frac{\partial P}{\partial z} \tag{3.8}$$

This describes the volume flow rate inside a cylinder in the laminar flow regime and is also known as the Hagen-Poseuille law. Similar analysis can be done for different geometrical shapes, however, a cylinder has symmetric properties which allow for a lot of simplifications of the Navier Stokes equations.

In many systems it is often good to gain an overview over the dimensionless numbers of the fluid such as the Reynolds number, the Prandtl number, and the Womersley number. These values allow us to gain some insight into the reigning forces in the fluid. An overview over the Reynolds and Prandtl numbers are shown in the Appendix chapter A.1 for blood and water.

3.1.2 Reynolds Number Re

The Reynolds number Re, gives an indication as to whether or not the flow is laminar or turbulent. It is expressed as the fraction of viscous to inertial forces in a liquid and is defined as:

$$Re = \frac{\rho v L_{ch}}{\mu} = \frac{\rho F L_{ch}}{\mu A} \tag{3.9}$$

with ρ [kg/m^3] as the liquid density, v [m/s] as the typical velocity in a pipe, e.g. the center-line velocity or an average velocity, L_{ch} [m] as the characteristic length scale determined as $L_{ch} = 2wb/(w + b)$ or $L_{ch} = \sqrt{wb}$ for a rectangular geometry (as per discussion in Appendix chapter A.1), and $L_{ch} = D$ for cylinders , μ [Pa s] as the dynamic viscosity, F [m^3/s] as the flow rate, and A [m^2] as the cross sectional area of the geometry. Re above 2300 is considered turbulent, and fully turbulent above 10000.

3.1.3 Prandtl Number

The Prandtl number Pr, is defined as the ratio between momentum diffusivity and thermal diffusivity.

$$Pr = \frac{c_p \mu}{k} \tag{3.10}$$

with c_p [J/(kg K)] as the specific heat, μ [Pa s] as the dynamic viscosity, and k [W/(mK)] as the thermal conductivity. Pr is constant for different liquids at set temperatures and in this case the values are therefore constant throughout the thesis (values shown in the Appendix in Table A.1) .

3.1.4 Womersley Number

Blood flow is pulsatile which introduces another complexity to be considered in the fluid dynamics. Fluid pulsatility can be described through the Womersley number:

$$\alpha = R\sqrt{\frac{\omega\rho}{\mu}} \tag{3.11}$$

With R [m] as the system radius, ω [1/s] as the frequency ($\omega = 2\pi f$), ρ [kg/m^3] as the density, and μ [Pa s] as the dynamic viscosity. A low Womersley number, $\alpha < 1$, indicates that the velocity profile has time to develop within the time of the frequency and a large velocity Womersley number, $\alpha > 10$, indicates a flat velocity profile.

In addition to these dimensionless numbers, there is one number which gives insight into the flow development along certain geometries. This number is the hydrodynamic entry length.

3.1.5 Hydrodynamic Entry Length

The hydrodynamic entrance length is defined as:

$$L_h = 0.05ReL_{ch} \tag{3.12}$$

with L_{ch} [m] as the characteristic length scale and Re as the Reynolds number.

The hydrodynamic entry length determines the distance from an inlet of a channel, L, a fluid with laminar flow will have to travel to actually assume the Hagen-Poiseuille shape. In the case of turbulent flow, the L_h is calculated as:

$$L_h = 10L_{ch} \tag{3.13}$$

3.2 Heat Transfer

3.2.1 In Solids

Similarly to the Navier Stokes equations expressing the conservation of momentum, mass flow and energy, one can express a heat equation from the energy conservation. The general heat transfer in solids can be described through as described in [117]:

$$\rho c_p(\frac{\partial T}{\partial t} + \mathbf{v}\nabla T) + \nabla \cdot (\mathbf{q} + \mathbf{q}_r) = -\alpha T : \frac{dS}{dt} + Q \tag{3.14}$$

With ρ as the solid density, c_p [J/(kg K)] as the specific heat capacity, T [K] as the temperature, v [m/s] as the velocity of translational motion, $\mathbf{q} = -k\nabla T$ [W/m^2] as the conductive heat flux, $\mathbf{q}_r$ [W/m^2] as the radiative heat flux, α [1/K] as the coefficient of thermal expansion, S [Pa] is the stress tensor, and Q [W/m^3] as the thermal power densities.

In the cases in this thesis, no movement of the solids ($\mathbf{v}$ = 0 m/s), no radiative heat flux ($\mathbf{q}_r$), and also no thermoelastic dissipation (Q_{ted}=$-\alpha\ T : \frac{dS}{dt}$) is considered. Therefore the equation simplifies to:

$$\rho c_p \frac{\partial T}{\partial t} + \nabla \cdot \mathbf{q} = Q \tag{3.15}$$

3.2.2 In Fluids

Similar to how the heat transfer is expressed in the solid, the heat transfer is expressed through this energy balance in fluids as described in [117]:

$$\rho c_p(\frac{\partial T}{\partial t} + \mathbf{v}\nabla T) + \nabla \cdot (\mathbf{q} + \mathbf{q}_r) = \alpha_p T(\frac{\partial P}{\partial t} + \mathbf{v}\nabla P) + \tau : \nabla \mathbf{v} + Q \tag{3.16}$$

With ρ [kg/m^3] as the solid density, c_p [J/(kg K)] as the specific heat capacity, T [K] as the temperature, $\mathbf{v}$ [m/s] as the velocity vector, $\mathbf{q} = -k\nabla T$ [W/m^2] as the conductive heat flux, α_p [1/K] as the coefficient of thermal expansion at a pressure, P [Pa] is the pressure, τ [Pa] is the viscous stress tensor, and Q [W/m^3] is the thermal power density other than viscous dissipation.

In these cases no radiative heat flux ($\mathbf{q}_r$), no thermal expansion due to the negligible temperature increases on the fluid properties ($\alpha_p T(\frac{\partial P}{\partial t})$, and also no viscous dissipation ($\tau : \nabla \mathbf{v}$) is considered. The influence of viscous dissipation in the systems considered in this thesis is discussed in the Appendix A.2. Therefore the equation simplifies to:

$$\rho c_p(\frac{\partial T}{\partial t} + \mathbf{v}\nabla T) + \nabla \cdot \mathbf{q} = -\alpha_p T \mathbf{v}\nabla P + Q \tag{3.17}$$

3.2.3 Bioheat Transfer

In biological systems, there are intrinsic sources of heat flow inside the solids considered. Specifically, the blood perfusion heats up the tissue and the metabolic thermal power density within the tissue is another source. This can be expressed as [117]:

$$\rho c_p \frac{\partial T}{\partial t} + \nabla \cdot \mathbf{q} = \rho_b c_{p,b} \omega_b (T_b - T) + Q_{met} \tag{3.18}$$

Where ρ [kg/m^3] is the tissue density, c_p [J/(kg K)] is the specific heat capacity of the tissue, T [K] is the tissue temperature, $\mathbf{q}$ [W/m^2] is the heat flux by conduction in the tissue, ρ_b [kg/m^3] is the blood density, $c_{p,b}$ [J/(kg K)] is the blood specific heat capacity, ω_b [1/s] is the blood perfusion rate, T_b [K] is the blood temperature, and Q_{met} [W/m^3] is the metabolic thermal power density of the tissue. As discussed in the state of the art, the blood perfusion term, $\rho c_{p,b} \omega_b (T_b - T)$, is one of the thermal power densities considered in infectious sites. However, in the thorax the tissue generally is intrinsically at the blood temperature T_b [K], meaning it is a source of heat flow in infectious sites at the peripherals. The immediate infectious region considered in this thesis is only the pathogens and the immune cells, and therefore not vascularized.

3.2.4 Conduction

There are three general types of heat transfer: conduction, convection and radiation. In the systems presented in this thesis, the main focus is on conduction and convection.

Conduction is the heat transfer between two solid interfaces and can be described through this equation:

$$q = -k\frac{\Delta T}{L} \tag{3.19}$$

Where q [W/m^2] is the heat flux, ΔT is the temperature difference ($\Delta T = T_2$ - T_1) between the two solid interfaces, L [m] is the distance between the two solid interfaces and the measured temperatures T_1 and T_2, and k [W/(mK)] is the thermal conductivity.

3.2.5 Convection

Convection describes the heat transfer between a fluid and a solid interface and can be described by the Nusselt number, Nu. The heat transfer at the boundary of the solid is set to be equal to the heat transfer of the fluid:

$$-k\frac{\partial T}{\partial z}(z=0) = h(T_w - T_\infty) \tag{3.20}$$

With z [m] as the distance between the solid and the liquid, k [W/(mK)] as the thermal conductivity of the solid, h [W/(m^{2}K)] as the heat transfer coefficient of the liquid, T_w [K] as the wall temperature, and T_∞ [K] as the bulk temperature of the liquid. The Nusselt number can is expressed as the ratio between the convective heat transfer and the conductive heat transfer.

$$Nu = -\frac{\partial(\frac{T_w - T}{T_w - T_\infty})}{\partial(\frac{z}{L})} = \frac{hL_{ch}}{k} \tag{3.21}$$

The Nusselt number relation gives an opportunity to determine the conductive heat transfer coefficient, h.

3.2.6 Thermal Entry Length

Similar to the hydrodynamic entry length for laminar fluids, L_h, there is a thermal entry length defined as:

$$L_{th} = 0.05RePrL_{ch} \tag{3.22}$$

If $Pr > 1$ then $L_{th} > L_h$ in all cases. In the case of blood at 37°C $Pr = 18.6$.

In the case of fully turbulent flow, both the hydrodynamic and thermodynamic entrance length are defined as:

$$L_h = L_{th} = 10L_{ch} \tag{3.23}$$

The knowledge of the L_h and L_{th} allows for the determination of the convective flow regime of the system as described in detail in the next section.

3.3 Theory for Aortic System and Microfluidic Chip

3.3.1 Flow in Aortic Channel

As previously mentioned, the geometry is simplified to a cylinder to calculate the flow conditions of the descending aorta. Figure 3.2 shows an overview over the different laminar flow regimes depending on L_h and L_{th} in a cylinder. As $Pr > 1$ in this case three different flow regimes are relevant; the undeveloped region $z < L_h, L_{th}$, the hydrodynamically developed region $z > L_h$ and $z < L_{th}$ and the fully developed region $z > L_h, L_{th}$. With z is the distance travelled from the inlet of the cylinder.

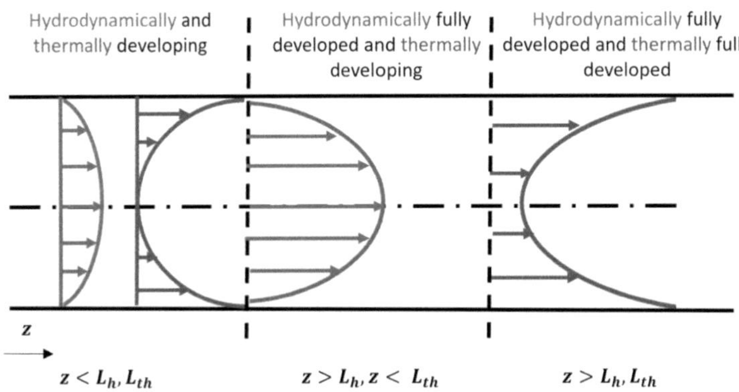

Figure 3.2: Flow regimes for the undeveloped region $z < L_h, L_{th}$, the hydrodynamically developed region $z > L_h$ and $z < L_{th}$ and the fully developed region $z > L_h, L_{th}$. Drawn based on literature [118].

Thus, it is possible to determine the different flow regimes in a cylinder with different radii, travel lengths and fluid velocities as shown in Figure A.1. These flow conditions determine the different equations which govern the flow conditions in the cylinder. An overview over different flow conditions for different cylinder radii is shown in the Appendix in chapter A.3.

3.3.2 Convective Flow Regimes in Cylinders

In the laminar flow regime ($Re < 2300$) there are three different flow regimes: when $z < L_h, L_{th}$ the fluid is in the undeveloped region, when $z > L_h$ and $z < L_{th}$ the fluid is in the hydrodynamically developed region (Graetz flow), and when $z > L_h, L_{th}$ the fluid is in the fully developed region. The knowledge of the flow conditions dictate the heat transfer between the fluid and the solid interface. The Nusselt number can be defined based on the flow condition. Thereby, the heat transfer coefficient, h, can be determined through the Nusselt number, Nu.

3.3.2.1 Convective Flow in Cylinder $z > L_h, L_{th}$: Fully Developed Flow

The velocity in the laminar fully developed flow region in a cylinder can be expressed with Hagen-Poiseuille as described in literature [118]:

$$v(r) = -\frac{1}{4\mu}(\frac{dP}{dz})R^2[1 - (\frac{r}{R})^2] \tag{3.24}$$

With P [Pa] as the fluid pressure, and μ [Pa s] as the dynamic viscosity, R [m] as the geometry radius, and v [m/s] as the axial component of the fluid velocity.

The mean velocity of the axial component can thus be determined as:

$$v_m = \frac{2}{R^2}\int_0^R v(r, z)rdr = -\frac{R^2}{8\mu}\frac{dP}{dz} \tag{3.25}$$

The average temperature, T_m, inside of the cylinder can be described as:

$$T_m = \frac{2}{v_m R^2}\int_0^R v(r, z)T(r)rdr \tag{3.26}$$

The fully developed temperature profile for a constant surface heat flux q_s is described in the Appendix by equation A.19. And by solving for the velocity and temperature profiles, h can be expressed as:

$$h = \frac{48}{11}(\frac{k}{D}) \tag{3.27}$$

In the fully developed region for laminar flow in a cylinder, the Nusselt number for constant surface heat flux can thus be determined as:

$$Nu_D = \frac{hD}{k} = 4.36 \tag{3.28}$$

with D as the cylinder diameter.

The detailed derivation is in the Appendix in Chapter A.4 and it is based on [118].

For isothermal conditions the Nusselt number can be determined as [118]:

$$Nu_D = \frac{hD}{k} = 3.66 \tag{3.29}$$

3.3.2.2 Convective Flow in Cylinder $z > L_h$ and $z < L_{th}$: Partly Developed Flow

For a constant surface temperature the Nusselt number for a thermally undeveloped and hydrodynamic developed flow regime is determined as [119] (Equation 4.13):

$$Nu_D = 1.62(RePr\frac{D}{z})^{1/3} \tag{3.30}$$

3.3.2.3 Convective Flow in Cylinder $z < L_h$ and $z < L_{th}$: Undeveloped Flow

In literature, the Nusselt number for a constant surface temperature in the hydrodnynamic and thermodynamic undeveloped flow regime in a cylinder is determined as [118] (Equation 8-62):

$$Nu_D = 3.66 + \frac{0.065(\frac{D}{z})RePr}{1 + 0.04[(\frac{D}{z})RePr]^{2/3}} \tag{3.31}$$

3.3.2.4 Flow in Cylinder Turbulent Transitional regime $2300 < Re < 8000$

In the low Reynolds number flow regime, the Nusselt number has been determined as in literature [119] (Equation 4.19):

$$Nu_D = 0.037(Re^{3/4} - 180)Pr^{0.42}[1 + (\frac{D}{z})^{2/3}] \tag{3.32}$$

3.3.3 Flow in Microfluidic Chip

In microfluidic chips with laminar flow, L_h and L_{th} are mostly small in relation to the total channel lengths. Therefore, in many cases the fluid is in the hydrodynamic and thermodynamic fully developed flow condition with $z >> L_h, L_{th}$. The heat transfer coefficient in microfluidic channels have been investigated in literature by Muzychka and Zovanovich [120]:

$$fRe_{\sqrt{A}} = \frac{12}{\sqrt{\epsilon}(1+\epsilon)[1 - \frac{192\epsilon}{\pi^5}\tanh(\frac{\pi}{2\epsilon})]} \tag{3.33}$$

With $fRe_{\sqrt{A}}$ as a defined friction-factor Reynolds number and ϵ = h/w as the aspect ratio of the channel with h and w as the height and width of the rectangular channel. Through this equation, it becomes clear that this friction-factor Reynolds number is merely dependent on the channel geometry in this flow region. To determine the Nusselt number, they use the following expression:

$$Nu_{\sqrt{A}} = C_1(\frac{fRe_{\sqrt{A}}}{8\sqrt{\pi}\epsilon^{\gamma}}) \tag{3.34}$$

With C_1 as an expression which is set to 3.86 for uniform wall flux and 3.24 for uniform wall temperature and γ as a shape parameter set to 1/10 for rounded or right-angled corners and -3/10 for angles with re-entrant corners. The heat transfer coefficient for the microfluidic chip is defined as:

$$h = \frac{kNu}{L_{ch}} \tag{3.35}$$

An important observation is that the heat transfer coefficient in microfluidic chip in the laminar fully developed hydrodynamic and thermodynamic flow region is only dependent on the geometry of the microfluidic channel. A theoretical determination of the heat transfer coefficient of a specific channel is shown in Figure 3.3.

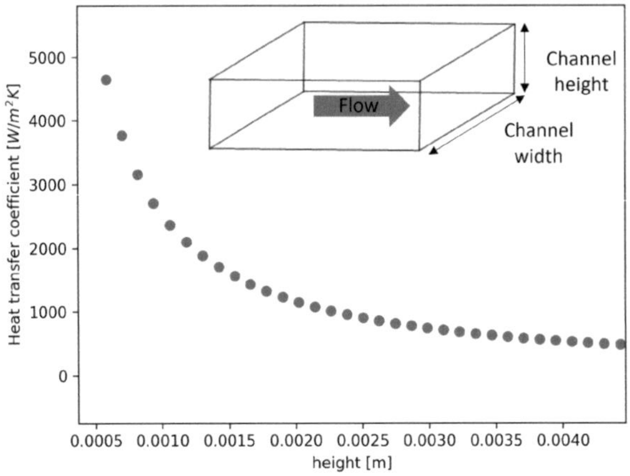

Figure 3.3: Theoretically determined heat transfer coefficient for the microfluidic chip with varying heights and determined width of 12 mm.

3.4 Thermal Lumped Element Model

Lumped element models allow for a simplified overview over the thermal pathways of a system. It is built like an electrical circuit where the voltage corresponds to a temperature difference, the current corresponds to heat flux, and the resistance corresponds to a calculated thermal resistance [121] as shown in Table 3.1. Each material in a thermal lumped element model can be described through resistors and capacitors. In this thesis a steady state system without any capacitance is considered. In a purely resistive circuit corresponding to steady state conduction, the conduction between two solid elements and convection between a solid and liquid interface is considered.

Table 3.1: Lumped Element Model parameters.

Electrical circuit and corresponding thermal element			
Electrical circuit	Unit	Thermal model	Unit
Voltage	V	Temperature	K
Current	A	Heat flow	W
Resistance	Ω	Thermal resistance	K/W

We consider two different types of heat transfer in this system: conduction or convection. In a lumped element model each element heat flows through can be represented by a resistance [121].

$$\mathrm{R}_{convection} = \frac{1}{h_{convection}\mathrm{A}_{surface}} \tag{3.36}$$

With $h_{convection}$ as the heat transfer coefficient and $\mathrm{A}_{surface}$ as the surface area through which the heat flow is conducted.

$$\mathrm{R}_{conduction} = \frac{l_{material}}{k_{material}\mathrm{A}_{surface}} \tag{3.37}$$

With $l_{material}$ [m] as the thickness of the material layer, $k_{material}$ [W/(mK)] as the thermal conductivity of the material, and $\mathrm{A}_{surface}$ [m^2] as the surface area of the material.

3.4.1 Lumped Element Model of Microfluidic Chips

3.4.1.1 One Level Microfluidic Chip

Parts of this chapter have been published in [122]. A microfluidic chip with a heat flux sensor embedded in PDMS above the microfluidic channel is considered. A lumped element model for this system is shown in Figure 3.4. The heat flow considered in this system is an exothermic chemical reaction expressed as $\dot{Q}_{source}$. In this system the heat flow is assumed to only have two directions; towards the sensor $\dot{Q}_{up}$ or down out of the microfluidic chip $\dot{Q}_{down}$ as illustrated in 3.4 a) and b). The considered surface area $A_{surface}$ (12 mm × 12 mm) is the width of the channel whereas the heat flux sensor area is smaller (10 mm × 10 mm). All of the parameters used are shown in the Appendix chapter B.1.

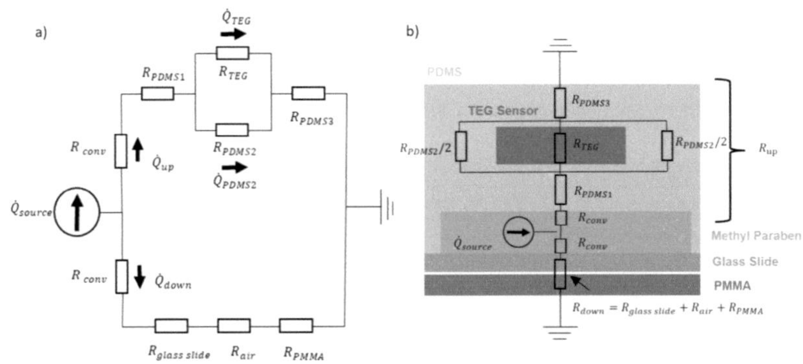

Figure 3.4: Lumped element model of the one level microfluidic chip. Open Access figure adapted from [122] published under open access under CC BY license (Creative Commons Attribution 4.0 International license).

The heat flow from the source initially passes a convective heat transfer element (R_{conv}) as shown in 3.4 a), which is the solid-fluid interface. Between the fluid and the heat flux sensor (TEG) there is a layer of PDMS represented as R_{PDMS1}. The heat flux sensor (R_{TEG}) is embedded in PDMS. The PDMS on both sides of the sensor is represented by R_{PDMS2} which is shown as a parallel pathway for the heat flow to flow in Figure 3.4 a). On top of the heat flux sensor there is another layer of PDMS R_{PDMS3}, above which there is a copper block which is assumed to be at the stable temperature of the environment (ground). In the direction down out of the microfluidic sensor, $\dot{Q}_{down}$; the glass slide of the microfluidic chip $R_{glassslide}$, a small air gap R_{air}, and the PMMA layer

R_{PMMA} are considered.

With the lumped element model as illustrated in Figure 3.4 a), current dividers are applied to express the source current $\dot{Q}_{source}$ in relation to the current (heat flow) flowing through the heat flux sensor $\dot{Q}_{TEG}$. This allows for an expression which describes the fraction of heat flow originating from the channel which gets detected by the heat flux sensor. The first current divider:

$$\dot{Q}_{up} = \dot{Q}_{source} \frac{\mathrm{R}_{down}}{\mathrm{R}_{down} + \mathrm{R}_{up}} \tag{3.38}$$

With a second current divider:

$$\dot{Q}_{TEG} = \dot{Q}_{up} \frac{\mathrm{R}_{PDMS2}}{\mathrm{R}_{PDMS2} + \mathrm{R}_{TEG}} \tag{3.39}$$

Inserting the expression for $\dot{Q}_{up}$ into the expression for $\dot{Q}_{TEG}$:

$$\dot{Q}_{TEG} = \dot{Q}_{source} \frac{\mathrm{R}_{down}}{\mathrm{R}_{down} + \mathrm{R}_{up}} \frac{\mathrm{R}_{PDMS2}}{\mathrm{R}_{PDMS2} + \mathrm{R}_{TEG}} \tag{3.40}$$

The two heat flows can thereby be expressed in relation to each other, also known as the heat transfer efficiency when expressed as a percentage:

$$\frac{\dot{Q}_{TEG}}{\dot{Q}_{source}} = \frac{\mathrm{R}_{down}}{\mathrm{R}_{down} + \mathrm{R}_{up}} \frac{\mathrm{R}_{PDMS2}}{\mathrm{R}_{PDMS2} + \mathrm{R}_{TEG}} \tag{3.41}$$

With $\dot{Q}_{TEG}$ as the heat flow through the sensor, $\dot{Q}_{source}$ as the exothermic heat flow originating from the microfluidic channel, $R_{down} = R_{conv} + \mathrm{R}_{glass} + \mathrm{R}_{air} + \mathrm{R}_{PMMA}$ as the sum of the material layers in series towards the bottom of the microfluidic chip, R_{up} as the layers of PDMS and embedded heat flux sensor as illustrated in Figure 3.4, R_{PDMS2} as the layers of PDMS around the sensor, and R_{TEG} as the heat flux sensor. The determined fraction of $\frac{\dot{Q}_{TEG}}{\dot{Q}_{source}}$ expressed as a percentage is 73% (the parameters used for the calculations are shown in the Appendix chapter B.1) .

3.4.1.2 Double Channel Microfluidic Chip

The double channel microfluidic chip presented in this section is based on a published system [123]. In this system, a second channel is added above the heat flux sensor (no copper block in this system). This is to further emulate the thermal environment of the aortic vascular graft by matching the heat transfer coefficient in the aorta with the heat transfer coefficient in the microfluidic channel. This system is a variation of the one level microfluidic chip presented above with slightly different chip parameters. The lumped element model for the double channel microfluidic chip is shown in Figure 3.5. The top resistance is the top channel and it is either filled with flowing water or empty (air). All of the parameters used are shown in the Appendix chapter B.2.

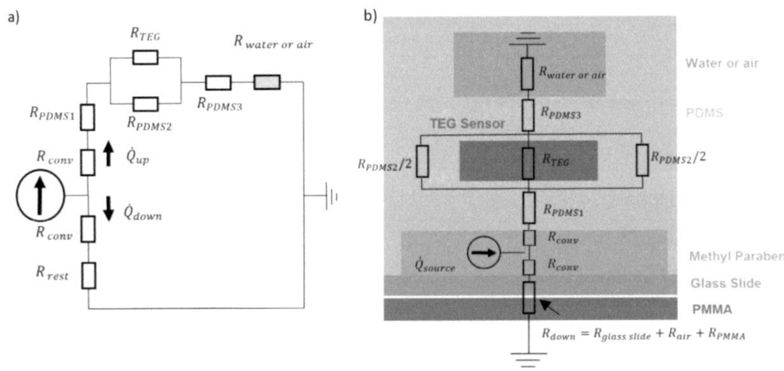

Figure 3.5: Lumped element model of the double channel microfluidic chip.

The equations for the lumped element model in this system are as follows:

$$R_{down} = R_{conv} + R_{glassslide} + R_{air} + R_{PMMA} \tag{3.42}$$

With the R_{conv} being the convective heat transfer as determined by the equation 3.36, and the rest determined by their conductive material properties as determined by the equation 3.37. In addition, $R_{PDMS1} = R_{PDMS3}$.

$$\mathrm{R}_{up} = \mathrm{R}_{conv} + 2\mathrm{R}_{PDMS1} + \frac{\mathrm{R}_{TEG}\mathrm{R}_{PDMS2}}{\mathrm{R}_{TEG} + \mathrm{R}_{PDMS2}} + \mathrm{R}_{\frac{w}{a}} \tag{3.43}$$

Where $\mathrm{R}_{\frac{w}{a}}$ as either the convective heat transfer for water flowing in the microfluidic channel or the conductive properties of air.

The heat flowing through the sensor in relation to the input heat flow is thus be expressed as:

$$\frac{\dot{Q}_{TEG}}{\dot{Q}_{source}} = \frac{\mathrm{R}_{down}}{\mathrm{R}_{down} + \mathrm{R}_{up}} \frac{\mathrm{R}_{PDMS2}}{\mathrm{R}_{PDMS2} + \mathrm{R}_{TEG}} \tag{3.44}$$

The determined fraction of $\frac{\dot{Q}_{TEG}}{\dot{Q}_{source}}$ expressed as a percentage is 77% with flowing water and 55% with air.

3.5 Finite Element Model Simulation Setup

3.5.1 COMSOL Fluid Dynamics

The following section is based on a previously published paper [1]. A descending aorta is simulated using a rectangular 2D axisymmetric simulation with a radius of 1 cm and a length of 20 cm as shown in Figure 3.6 a). The simulation is performed using COMSOL Multiphysics 6.0 (COMSOL Inc., Sweden). In the simulation the following assumptions are used: no Windkessel effect, blood is approximated as a Newtonian fluid, and a straight aorta with no branching.

As described in chapter 3.1.2, the Reynolds number defines the fluid flow region. The Reynolds number of the blood flow in an aorta is on average in the low Reynolds number turbulent regime ($2300 < Re < 8000$) also called the translational flow region. Furthermore, the different types of flow dependent on the hydrodynamic L_h and thermal L_{th} entry length can be defined as described in chapter 3.3.2.

The simulation build is shown schematically in Figure 3.6 b). The fluid flow simulation is first initialized by a wall distance initialization with the low Re $k - \epsilon$ turbulent flow module (from COMSOL) as shown in Step 1 in Figure 3.6 b). Step 2 uses the isothermal flow Multiphysics feature connecting the low Re $k - \epsilon$ turbulent flow module (spf) and heat transfer in solid and fluids module (ht). Its starting point for the simulation is the solution of Step 1.

Study 2 (Figure 3.6 b)) shows the time-dependent study. The solutions of the stationary study is used as the initial condition of the time-dependent study. Similar to the stationary study, the simulation starts with a wall distance initialization (Step 1). It continues (Step 2) with a partial differential equation (PDE) which describes the pulsatile flow and the isothermal flow Multiphysics feature using both the low Re $k - \epsilon$ turbulent flow module (spf) and heat transfer in solid and fluids module (ht). The time-dependent system describes a pulsatile flow in the axial direction represented by the equation:

$$\mathrm{v}_z = \mathrm{v}_{avg} + \mathrm{u} \tag{3.45}$$

Where v_z [m/s] is the velocity in the axial direction, v_{avg} [m/s] is the average velocity of blood flow, and u [m/s] is the pulsatile element of flow.

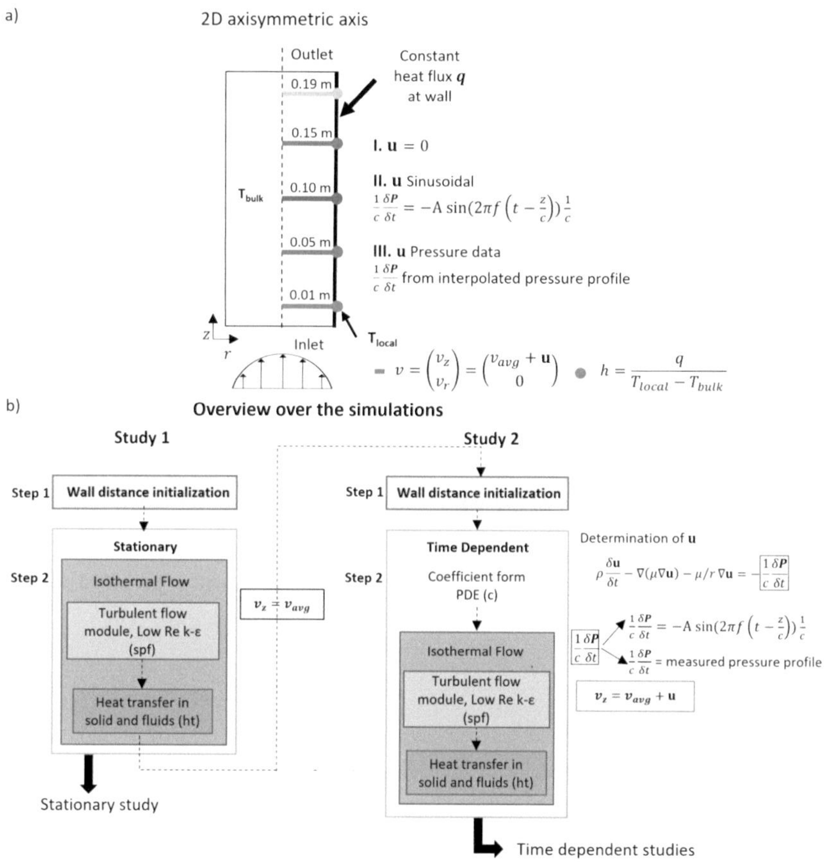

Figure 3.6: a) Sketch overview over the flow simulation. b) Schematic overview over the detailed simulation build for both the stationary and the time dependent simulation. Open Access figure from [1] published under open access under CC BY license (Creative Commons Attribution 4.0 International license).

The pulsatile flow equation describing the Womersley flow is expressed using a general coefficient form PDE as shown in the Appendix in chapter A.5 and is as follows:

$$\rho\frac{\partial \mathrm{u}}{\partial t} + \nabla(-\mu\nabla \mathrm{u}) - \mu/r\nabla \mathrm{u} = -\frac{1}{c}\frac{\partial P}{\partial t} \tag{3.46}$$

Different types of pulsatile flow can be introduced through the source term $\frac{1}{c}\frac{\partial P}{\partial t}$. For example a sinusoidal pulsatility:

$$\frac{1}{c}\frac{\partial P}{\partial t} = -Asin(2\pi f(t - \frac{z}{c}))\frac{1}{c} \tag{3.47}$$

Where P [Pa] is the pressure, c is the pulse wave velocity 8.8 m/s, A [Pa/s] as the amplitude, and f [Hz] as the frequency.

One can also use a measured pressure from literature as an input pulsatility [124] .

3.5.2 COMSOL Thermodynamics

The Heat Transfer in Solids and Fluids (ht) module is used to introduce the heat transfer aspect in the simulation. The simulation model is a cross-section of an aorta as shown in Figure 3.7 a). The range of heat transfer coefficients, h, defined in the fluid dynamics simulations is applied in the thermodynamic simulation. For a stationary heat transfer study, the heat transfer equation used is:

$$\rho c_p \mathrm{v} \nabla T + \nabla \cdot \mathbf{q} = Q + Q_{ted} \tag{3.48}$$

With $q = -k\nabla T$ as the heat flux and $Q_{ted} = -\alpha T : \frac{dS}{dt}$ as the thermoelastic dissipation contribution in the time dependent case:

$$\rho c_p \frac{\partial T}{\partial t} + \rho c_p \mathrm{v} \nabla T + \nabla \cdot \mathbf{q} = Q + Q_{ted} \tag{3.49}$$

With $\mathbf{q} = -k\nabla T$. A general inward heat flux of $q_{in} = 200$ W/m^2 is applied at the boundary of $R = 0.01$ m.

The heat transfer coefficient at the cylinder wall is determined by:

$$h = \frac{q}{T_{local} - T_{bulk}} \tag{3.50}$$

with the given constant heat flux at the wall q, the local temperature at the wall T_{local} and the bulk temperature:

$$T_{bulk} = \frac{\int_0^{r/2} 2\pi \mathrm{v}(r,z) T(r,z) dr}{\int_0^{r/2} 2\pi \mathrm{v}(r,z) dr} \tag{3.51}$$

Both Study 1 and Study 2 are initialized using the wall distance initialization, whereafter the isothermal flow module is used to solve for the turbulent flow in the low Re flow region in combination with the heat transfer. In all of the simulations, the boundary conditions are set to 310.15 K (except the boundary at which the heat transfer coefficient is set) due to the minimal expected increase in temperature.

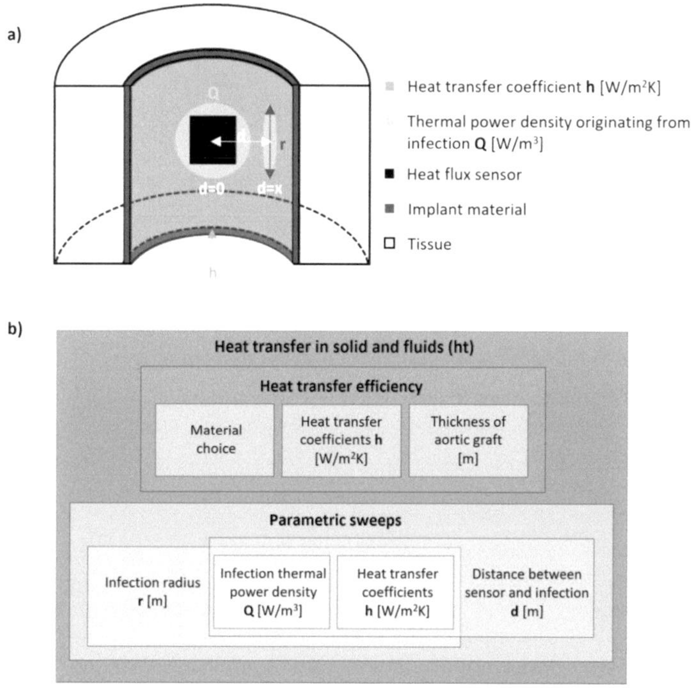

Figure 3.7: Simulation overview over the thermal simulation for the infection detection vascular graft system. Open Access figure from [1] published under open access under CC BY license (Creative Commons Attribution 4.0 International license).

Different parametric variations (material choice, heat transfer coefficient, aortic graft thickness, infection radius, infectious thermal power density, and distance between the sensor and the infection) are investigated in the thermodynamic simulations as shown in Figure 3.7 b).

4 Fabrication and Methods

4.1 Microfluidic Chip Fabrication

The following subsections are based on previously published work [1, 122, 123, 125, 126]. They detail the bottom-up fabrication steps of the microfluidic chips with integrated heat flux sensors. The specific fabrication steps cover also the lithography steps (chapter 4.1.1.1) for the preparation of the silicon molds used for the soft lithography steps (chapter 4.1.1.2).

Further, the three experimental systems (as shown in Figure 4.1) and experimental procedures are discussed. The microfluidic chips can be grouped in the "one level microfluidic chip" (chapter 4.1.1) and the "double channel microfluidic chip" (chapter 4.1.2). Furthermore there are exothermic chemical reaction experiments (chapter 4.2.1) and bacterial experiments (chapter 4.2.2). A color version of this chapter is available at https://doi.org/10.3929/ethz-b-000703739.

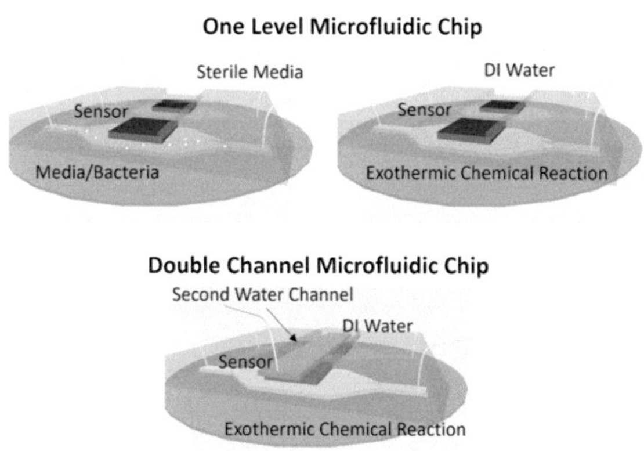

Figure 4.1: Overview over the different microfluidic chips fabricated and used in chapters 5.1.2.1, 5.1.2.3, 5.2, and 5.3, and published previously in literature [122, 123, 125, 126]. Parts of this is an Open Access figure adapted from [125] published under open access under CC BY license (Creative Commons Attribution 4.0 International license).

4.1.1 One Level Microfluidic Chip Fabrication

4.1.1.1 Lithography Steps

The designed microfluidic channel structures are drawn with KLayout and transferred to a 5-inch soda lime mask for lithography using a DWL 2000 laser pattern generator. The soda lime mask is then etched in Chrome ETCH N1 for 40 s to 50 s. Subsequently, the mask is rinsed and then placed in Technistrip P1316 for 10 min. This mask is used in the lithography fabrication steps as shown in Figure 4.2.

The specific lithography fabrication steps are described further in detail according to the order illustrated in Figure 4.2.

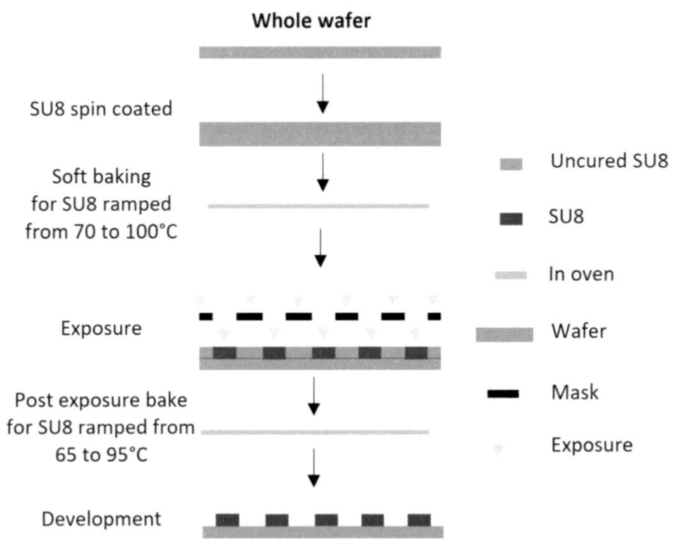

Figure 4.2: Lithography steps for the silicon wafer coated with SU8. This wafer with SU8 structures is the mold for the further soft lithography steps for the fabrication of the microfluidic chip. Open Access figure adapted from [125] published under open access under CC BY license (Creative Commons Attribution 4.0 International license).

As a first step, a 4 inch silicon wafer is spin coated with 4 mL of SU8 with the respective spin coating parameters as listed in Table 4.1.

Table 4.1: Spin coating parameters for different SU8 thicknesses and types.

Spin coating parameters			
Step	Time	Velocity	Acceleration
SU8 2150 with 320 µm			
Step 1	30 s	500 rpm	200 rpm/s
Step 2	30 s	1000 rpm	300 rpm/s
SU8 100 with 170 µm			
Step 1	10 s	500 rpm	200 rpm/s
Step 2	60 s	1250 rpm	300 rpm/s

Following the spin coating step, the wafer with uncured SU8 is heated on a hot plate. Depending on the thickness of the SU8 (320 µm or 170 µm), the time of the soft bake is tuned, as shown in Table 4.2. At this step, it is of importance that the SU8 is not sticky when it is cooled down to room temperature before proceeding to the exposure step. Furthermore, thick SU8 layers over 200 µm could crack if the temperature is not ramped up slowly.

Table 4.2: Soft baking parameters for different SU8 thicknesses.

Soft Bake Parameters			
Step	Time	Temperature	Comment
SU8 2150 with 320 µm			
Step 1	9 min	70°C	Wafer placed on hot plate
Step 2	>10 min	70°C to 100°C	Be aware of cracking
Step 3	97 min	100°C	Baking step
Step 4	15 min	75°C	Cool down
Step 5	Indefinite	RT	Cool down
SU8 100 with 170 µm			
Step 1	70 min	95°C	Wafer placed on hot plate
Step 2	Indefinite	RT	Cool down
In case SU8 is soft after cool down			
Step 1	10 min	70°C	Wafer placed on hot plate
Step 2	>10 min	70°C to 100°C	Be aware of cracking
Step 3	15 min	100°C	Baking step

Following the soft baking step and ensuring that the SU8 is not soft at room temperature, the wafer is exposed using a Karl Suess Mask Alginer (MA6). The parameters are shown in Table 4.3.

Table 4.3: Exposure parameters for the different SU8 thicknesses.

Exposure Parameters		
Time	Intensity	Energy
SU8 2150 with 320 µm		
66 s	440 mW/cm^2	400 mJ/cm^2
SU8 100 with 170 µm		
51 s	400 mW/cm^2	400 mJ/cm^2

After the exposure, the wafers are then post exposure baked depending on the SU8 thickness as shown in Table 4.4.

Table 4.4: Post exposure baking parameters for different SU8 thicknesses.

Post Exposure Bake Parameters			
Step	Time	Temperature	Comment
SU8 2150 with 320 µm			
Step 1	5 min	65°C	Wafer placed on hot plate
Step 2	>5 min	65°C to 95°C	Be aware of cracking
Step 3	25 min	95°C	Baking step
Step 4	Indefinite	RT	Cool down
SU8 100 with 170 µm			
Step 1	16 min	95°C	Wafer placed on hot plate
Step 2	Indefinite	RT	Cool down

The wafer is then developed in Mr Dev 600 (micro resist technology GmbH) for the given time as shown in Table 4.5.

Table 4.5: Development parameters for the different SU8 thicknesses.

Development Parameters		
Step	Time	Comment
SU8 2150 with 320 µm		
Step 1	25 min	In Mr Dev 600
Step 2	1 min	In fresh Mr Dev 600
SU8 100 with 170 µm		
Step 1	16 min	In Mr Dev 600
Step 2	1 min	In fresh Mr Dev 600

The wafer is rinsed in isopropanol and in deionized water. As a last step, the wafer with SU8 is baked at 150°C for 5 min. The fabricated structures are diced in the appropriate pieces.

4.1.1.2 Soft Lithography Steps

The diced wafer with the microfluidic channel structures is then used for soft lithography for the fabrication of the microfluidic chip as shown step-by-step in Figure 4.3.

First, PDMS (Sylgard 184) is mixed for 4 min in a 1:10 ratio between elastomer and crosslinker. The mixture is subsequently vacuum desiccated for 40 min or until there are no more air bubbles. The diced wafer with the SU8 microstructures is placed in a petri dish and PDMS is poured over the structures. This is again vacuum desiccated for 40 min or until there are no air bubbles. Using spacers (metal pieces with a defined thickness), a layer of 0.15 mm PDMS is defined. The PDMS is then cured at 80°C for 3 h.

After the first PDMS curing step, the heat flux sensors are fixed directly on top of the PDMS. Uncured PDMS is poured over the fixed heat flux sensors and a defined thickness of 150 µm is again controlled with a spacer on top. The chip is then vacuum desiccated for 40 min and cured for 80°C for 3 h.

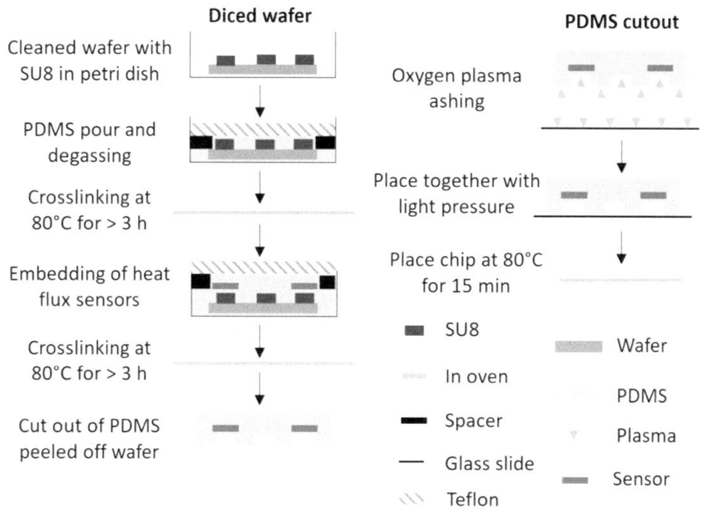

Figure 4.3: Soft lithography steps for the fabrication of one level microfluidic chip. Open Access figure adapted from [125] published under open access under CC BY license (Creative Commons Attribution 4.0 International license).

The cured PDMS with the embedded heat flux sensors is carefully cut and lifted from the silicon wafer. A 20 G needle is used to punch out the inlet and outlet channels and the cured PDMS is fused to a glass slide using oxygen plasma (at 0.8 mbar with 100 W in O_2 for 90 s). The chip is then heated in the oven for 15 min at 80°C for increased adhesion to the glass slide.

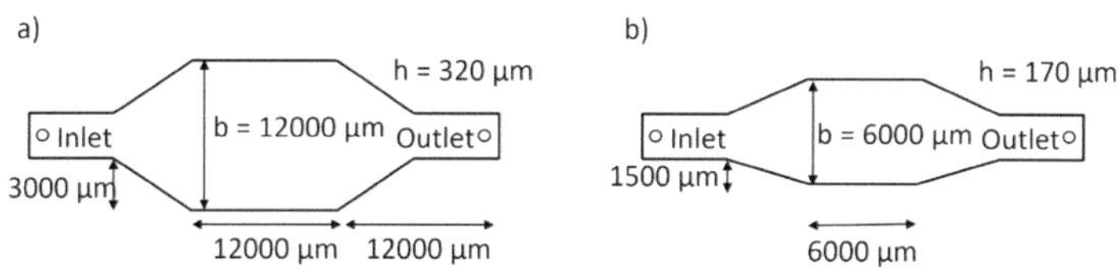

Figure 4.4: The geometry of the fabricated chips of different thicknesses.

The two different channel thicknesses also represent two different channel geometries used in the different experiments in the experimental chapter 5. Figure 4.4 a) shows the microfluidic chip geometry used in chapter 5 and 5.1.2.3, and Figure 4.4 b) shows the microfluidic chip geometry used in chapters 5.2 and 5.3.

4.1.2 Double Channel Microfluidic Chip Fabrication

The double channel microfluidic chips have an additional water channel at the top of the heat flux sensor for the purpose of imitating the thermal environment of the infected vascular graft system. The preparation of the top and bottom parts of the microfluidic chip are sketched in Figure 4.5.

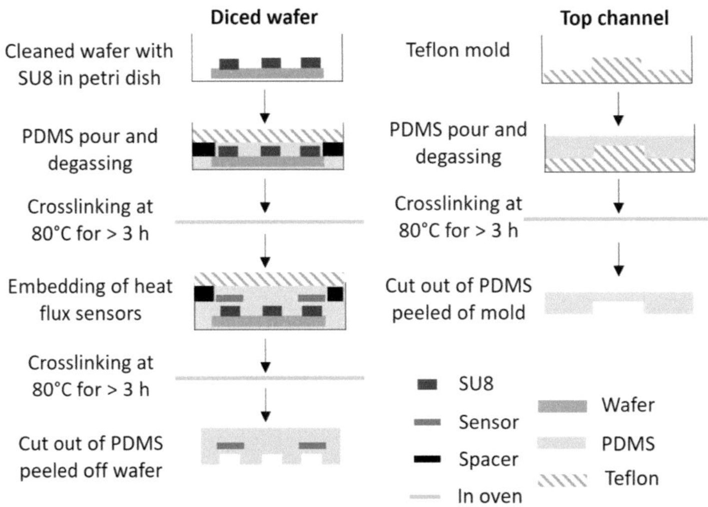

Figure 4.5: Schematic of the preparation of the bottom and top channel for the double channel microfluidic chip. Open Access figure from [125] published under open access under CC BY license (Creative Commons Attribution 4.0 International license).

The bottom channel can be fabricated with different thicknesses as discussed in the one level microfluidic chip in chapter 4.1.1. Similar to the bottom channel fabrication steps, the fabrication for the top channel starts with vigorously stirring the PDMS for 4 min in a 1:10 ratio between the elastomer and crosslinker. This mixture is vacuum desiccated for 40 min and poured on top of the teflon mold with the desired geometry. The teflon mold is pre-fabricated in the desired shape. The petri dish containing the teflon mold and the uncured PDMS is also vacuum desiccated for 40 min or until there are no air bubbles. The PDMS is cured at 80°C for 3 h in an oven and subsequently cut out of the mold.

As shown in Figure 4.6, the bottom channel and top channel parts are fused together on top of a glass slide using oxygen plasma ashing. The oxygen plasma applied is for 1 min and 30 s at 100 W in an O_2 environment. Subsequently, the assembled chip is placed in the oven for 15 min at 80°C for maximal adhesion between the layers.

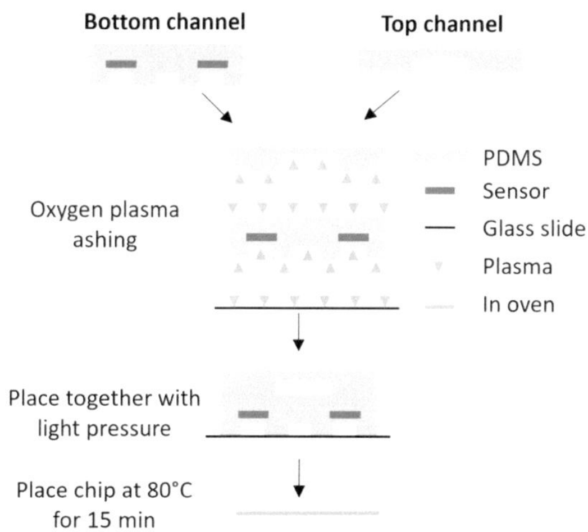

Figure 4.6: Schematic of the assembly of the bottom and top channel with the glass slide using oxygen plasma ashing.

4.2 Experimental Setup

Different experimental setups are used to measure the microfluidic chips. The biological environment around the aorta is relatively stable at 37°C, therefore the experimental environment for the *in vitro* microfluidic chips is also 37°C. Three different 37°C setups are used; oven, incubator, and a temperature controlled room. In principle, the different systems function similarly which is described in detail for the different experiments. An exothermic chemical reaction is used to thermally calibrate the chip and a biological sample is placed in the chip to measure its metabolic growth. All of these experimental systems will be described in this chapter.

4.2.1 Exothermic Chemical Reaction Calibration Setup

Each microfluidic chip has two channels, one for measuring and compensating for the environmental thermal fluctuations and one channel containing the sample to be measured. The experimental procedure of the exothermic chemical reaction in the microfluidic chip is set up as shown in Figure 4.7. At the beginning of each experiment, the chip and all the equipment is placed in the 37°C environment and thermally stabilized for multiple hours or even overnight.

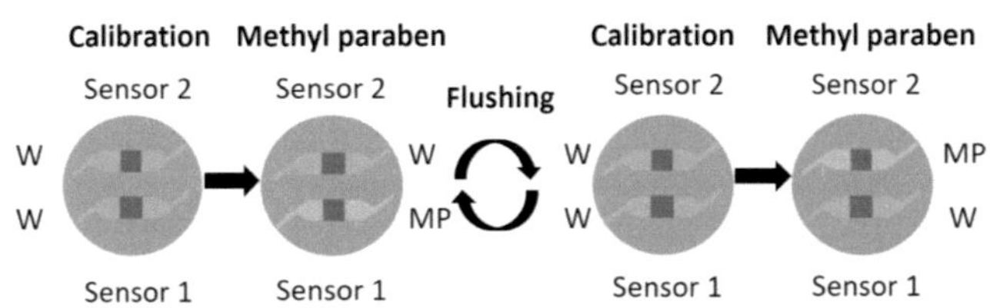

Figure 4.7: Experimental procedure for the exothermic chemical reaction in microfluidic chips. Open Access figure from [122] published under open access under CC BY license (Creative Commons Attribution 4.0 International license).

Initially, DI water is flowing in both channels as shown in Figure 4.7. The experiment is performed according to exothermic chemical calibration experiments for flowing microcalorimeters as previously shown in literature [127]. 0.5 M of NaOH (Sigma-Aldrich) is mixed with 152.32 mg methyl-4-hydroxy-benzoat, methyl paraben, (Sigma-Aldrich $\geq$ 99.0% *ReagentPlus*). Immediately after mixing, the exact time is tracked and the fluid is flown through one of the channels ($t = 0$ s of the experiment). After this experiment, the channels are flushed with DI water and the experiment is repeated again with switched channels as illustrated in the Figure 4.7.

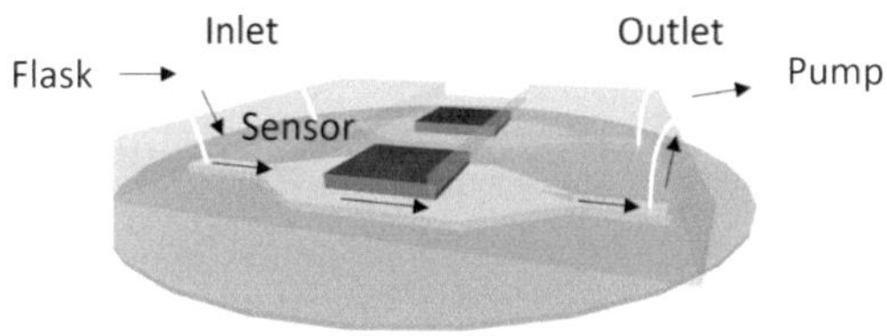

Figure 4.8: Overview over the inlet and outlet of the microfluidic chip.

A sketch of the microfluidic chip is shown in Figure 4.8. PTFE tubing (Cole Palmer 06417-11) is used to transport the liquids from the flasks to the channels. The inner diameter of these tubes is 0.012 inches and the distance of the tubing is 0.3 m. The PTFE tubing is inserted into the peristaltic tubing with an inner diameter of 0.020 inches (P720/TS-020S Instech Laboratories).

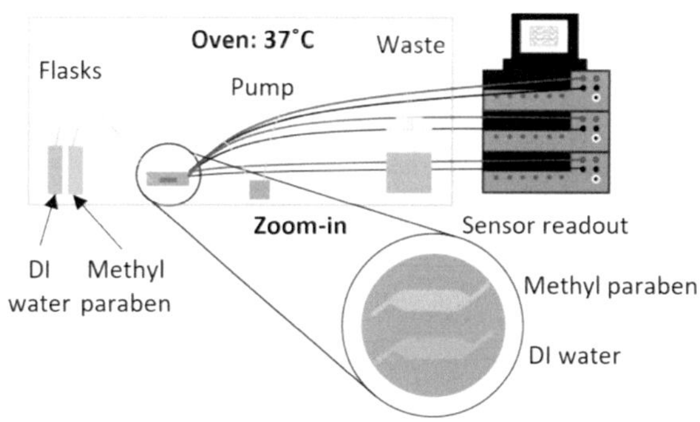

Figure 4.9: Microfluidic experimental setup of the methyl paraben system in temperature regulated oven. Open Access figure from [122] published under open access under CC BY license (Creative Commons Attribution 4.0 International license).

Figure 4.9 shows the experimental setup of the methyl paraben calibration experiment. An oven (Voetsch 4006) at 37°C is used for the thermal stable environment. A peristaltic pump (Instech P720) is used to control the flow rate from the containers with DI water and with the methyl paraben mixed with NaOH. All of the equipment including the microfluidic chip is placed inside the oven prior to the data acquisition. The two heat flux sensors, greenTEG gSKIN XP 26 C (datasheet in Appendix chapter C.1), embedded in the microfluidic chip are connected to two Keithly 2000 multimeters to measure the voltage. A temperature sensor, also embedded in PDMS is measured using a Keithly 2400 sourcemeter and the temperature value is directly applied to convert the measured voltage to heat flux as according to the given datasheet by greenTEG. These sensors are measured using the python package 'visa' from 'pyvisa' for control, data acquisition, and real-time readout.

4.2.2 Microfluidic Bacterial Experimental Setup

The experiment with growing bacteria is conducted similarly to the methyl paraben thermal calibration experiment described in the previous section as illustrated in Figure 4.10. Both channels are initially filled with sterile lysogeny broth (LB), also referred to as media. One of the containers with media (50 mL) is then filled with 1 mL of MG1655 *Escherichia Coli* (*E. coli*) which has been growing overnight. After the experiment, both channels are sterilized with 70% ethanol (EtOH) mixed with sterile water. The experiment is then repeated with bacteria in the other channel.

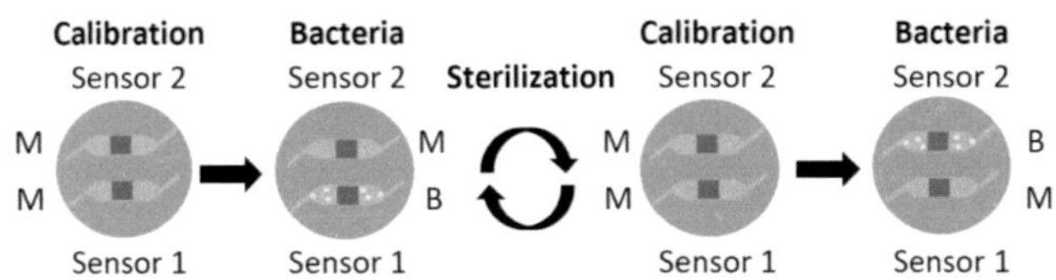

Figure 4.10: Microfluidic experimental setup of the bacterial experimental system in the incubator. Open Access figure from [122] published under open access under CC BY license (Creative Commons Attribution 4.0 International license).

Two different systems, as also compared in detail in a publication [126], are discussed in the following subsections. A thermally stabilized experimental setup and a thermally unstable experimental setup are presented.

4.2.2.1 Thermally Stable Environment

In preparation for the experiment, *E. coli* is grown in LB media overnight at 37°C in an incubator from a single colony on a petri dish. All of the equipment is sterilized and placed inside the sterilized incubator the day before the experiment. As shown in Figure 4.11, the microfluidic chip is placed inside of a thermally stabilizing polymethyl methacrylate (PMMA) box, with two 4 mm PMMA layers separated by a 12 mm layer of air in between. To further ensure thermal stability, a copper piece is placed on top of the microfluidic chip.

Differently to the methyl paraben experiment, overall sterility and the prevention of bacterial adhesion in the tubing is of importance in the experiments measuring bacterial growth. To ascertain the sterility of the LB media throughout the experiment, a sterile needle (21G Sterican) is attached to a 0.22 µm filter (Millipore) is inserted through the lid of both of the flasks. The sterile needle without the filter is inserted into the PTFE tubing (Cole Palmer 06417-11) for the contents of the flasks to be transported out without compromising the flask sterility. The needle with the filter allows for sterile air to enter the flask and to fill up the removed volume. To prevent bacterial adhesion in the tubing wall a 1:100 ratio of tween20 is added to the LB media in which *E. coli* will grow.

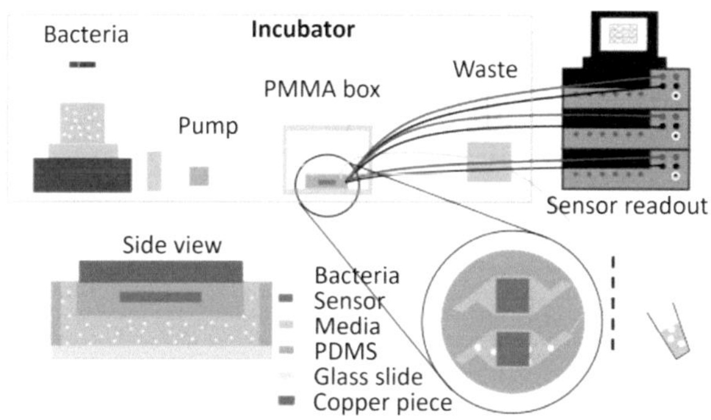

Figure 4.11: Microfluidic experiment in an incubator. Pictures of this system is shown in the Appendix in chapter D.1. Open Access figure from [122] published under open access under CC BY license (Creative Commons Attribution 4.0 International license).

The bacteria is grown in a flask using a magnetic stirrer at 200 rpm to ensure optimal bacterial growth. To quantify the bacterial population, samples are collected every 25 min at the outlet of the microfluidic chip and measured using an optical density meter (OD meter).

The data acquisition is performed identically to the methyl paraben exothermic chemical reaction.

4.2.2.2 Thermally not Stabilized Environment

The experimental setup shown in Figure 4.12 is thermally less stable than Figure 4.11. The experiment is in a temperature controlled room and the microfluidic chip is placed in the room without a thermally stabilizing PMMA box or an additional piece of copper.

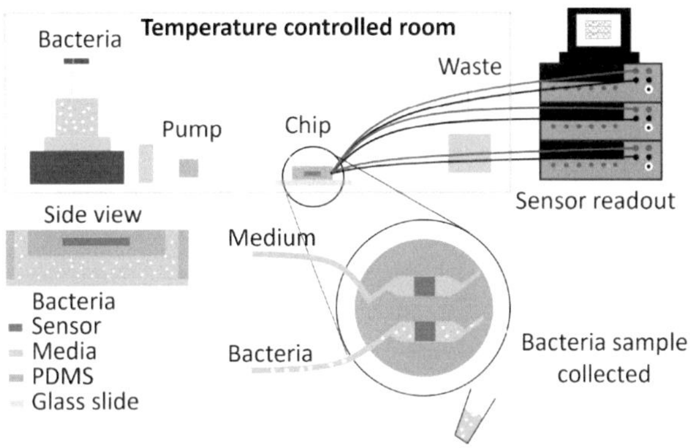

Figure 4.12: Microfluidic experiment with bacteria without thermal stabilization. Open Access figure from [125] published under open access under CC BY license (Creative Commons Attribution 4.0 International license).

Pictures of both the thermally stable environment (chapter 4.2.2.1) and the thermally not stabilized environments are shown in the Appendix in chapter D.1.

4.3 Microcalorimetric Experiments

Microcalorimetric experiments allow for an industry reference measurement of the total heat flow originating from a sample (chemical or biological). The following described experimental setup is at Basel University at the Department of Biomedical Engineering. These measurements were performed in collaboration with Dr. Olivier Braissant at the Department of Biomedical Engineering at the University of Basel.

4.3.1 Methyl Paraben Experiment

Directly upon mixing of 10 mL 0.5 M NaOH (Sigma Aldrich) and 76.66 mg of methyl paraben (Sigma Aldrich 99% ReagentPlus), aliquots of the samples are placed inside of the microcalorimeter (TAMII nano Water/TA, Delaware, USA) which is already pre-heated to 37°C. Small samples in the range between 20 µL and 200 µL can be filled inside the glass bottle as shown in Figure 4.13 and are slowly lifted into the respective cylindrical hole. Each exothermic sample is placed together with a reference sample containing the same amount of deionized water.

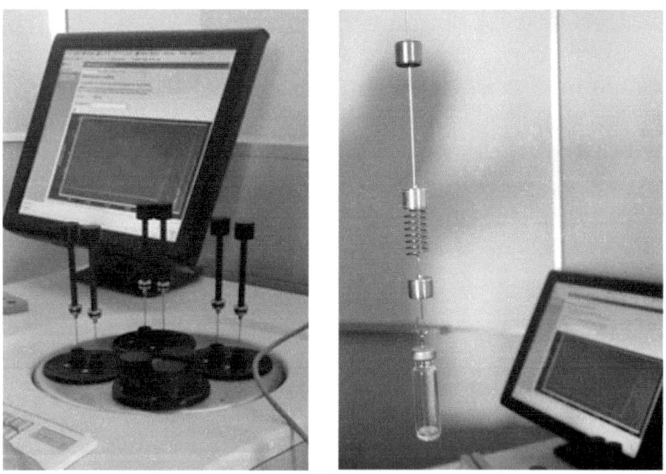

Figure 4.13: Picture of the TAM II nano microcalorimeter used for the measurement of the exothermic reaction of methyl paraben.

4.3.2 Immune Cell Experiment

Upon preparing both bacteria and immune cells to the point at which they are to be measured, 250 µL of the prepared media mixture with cytokines and cells are loaded within each well as shown in Figure 4.14. The top and bottom row of the plate is filled with sterile deionized water as a reference. Upon preparation and securing of the individual wells, the plate is placed into the microcalorimeter, Symcel Calscreener (Symcel Sverige AB), for an initial stabilization of the temperature at 37°C and the subsequent measurement. The samples are prepared within a sterile hood to prevent contamination and each individual well is sealed using a screwdriver to avoid evaporation and cross-contamination of the samples over the experimental period. Biological samples are typically left for 1-2 weeks within the microcalorimeter.

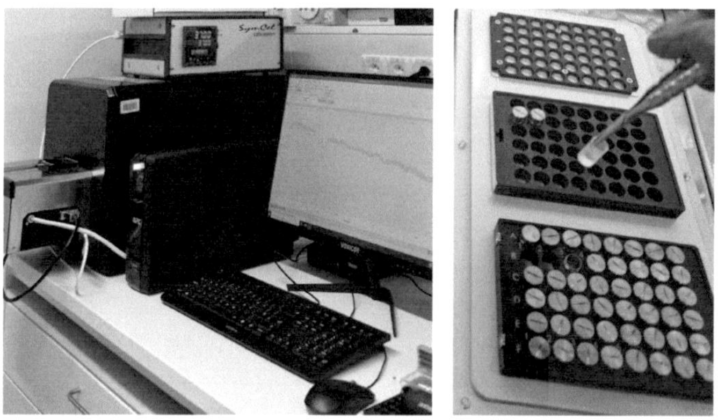

Figure 4.14: Picture of the Symcel Calscreener (Symcel Sverige AB) microcalorimeter used for the measurement of the biological samples.

4.4 Biological Protocols

As already introduced above, both bacteria and immune cells are used for experiments. Specifically, MG1655 *Escherichia Coli* (*E. coli*) and *Enterococcus Faecalis* (*E. faecalis*) are used to represent gram negative and gram positive bacterial populations, respectively. The laboratory used (Laboratory for Mucosal Immunology at ETHZ) for the culturing of these bacteria is of biosafetly level 2 (BSL2).

4.4.1 Bacterial Protocols: *Escherichia Coli* and *Enterococcus Faecalis*

Both strains are kept on LB agar plates at 4°C until they are used for further growth. The bacterial populations are grown overnight in LB media in an incubator at 37°C with a shaker to ensure oxygenation. Both strains are successfully grown this way.

Table 4.6: Chemicals a used per 1 L of LB media.

Lysogeny Broth Media	
Component	Amount
Tryptone	10 g
Yeast extract	5 g
NaCl	5 g

4.4.1.1 Bacterial Vaccine Protocol

Vaccine protocols were used for *E. coli* and *E. faecalis* bacteria to have access to the bacteria with their structural integrity intact without further bacterial growth. Treating live bacteria with a certain concentration of peracetic acid kills the bacteria without breaking the cell wall. These cells are subsequently used to stimulate the immune cells for microcalorimetry or FACS measurements. As the bacteria are dead, they do not contribute to the heat flow measured and their sole function is to stimulate the immune cells. The following vaccine protocol listed step-by-step in Table 4.7 is adapted from a protocol used in the Mucosal Immunology Laboratory in the Institute of Food Nutrition and Health in (D-HEST) at ETHZ.

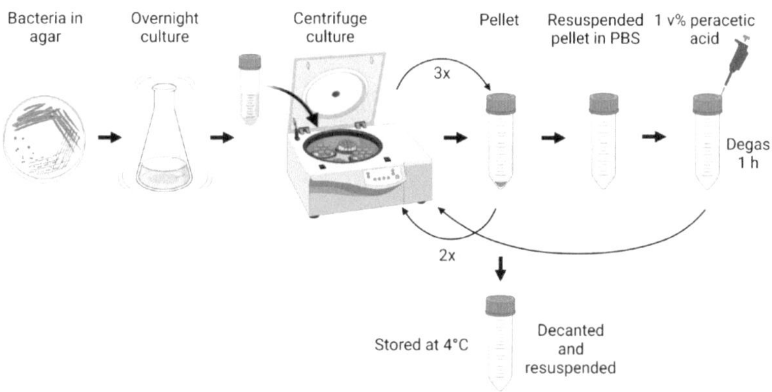

Figure 4.15: Illustration of the bacterial vaccine protocol. Sketch created with BioRender.com.

For specific bacterial concentrations, the bacterial vaccine concentration is measured using an optical density meter and diluted using phosphor buffered saline. The vaccine is stored at 4°C for a maximum of 6 weeks.

Table 4.7: Detailed description of the bacterial vaccine protocol used for *E. coli* and *E. faecalis*.

Bacterial Vaccine Protocol	
Step	Description
1	Grow appropriate amount of bacterial culture in LB media overnight in Erlenmeyer flask
2	Transfer bacterial culture to 1 L centrifuge beaker
3	Centrifuge at 7000 rpm for 15 min at 10°C
4	Dissolve pellets in 10 mL sterile PBS
5	Transfer to a clean 150 mL centrifuge beaker and make sure that it does not have any residual clumps
6	Add 1% (v/v) of peracetic acid (36-40%) to the volume of the pellet with the PBS
7	Make sure all of the beaker surface is touched by the peracetic acid making sure there are no live bacteria left
8	Slightly open the lid allowing for degassing and leave for 60 min at room temperature making sure the peracetic acid is decomposed to acetic acid
9	Add 50 mL sterile PBS
10	Centrifuge at 10 000 rpm for 10 min at 10°C
11	Resuspend in 10 mL sterile PBS
12	Repeat steps 9 to 11 two times
13	Decant and add PBS so that for each 50 mL initial culture, the final volume is 200 µL for a final concentration of 10^{10}-10^{11} CFU/mL
14	Dissolve the pellet using a shaker
15	Store the vaccine at 4°C for up to 6 weeks
16	Directly make a negative culture from the vaccine to make sure it is sterile

4.5 Immune Cell: THP-1

As previously discussed in the state of the art (chapter 2.4), THP-1 monocytes are considered as a representative model system for the monocyte-macrophage differentiation [67]. Furthermore, it is also considered as an *in vitro* system for investigations of immune cell/pathogen interaction. The THP-1 monocyte cells are grown in different media to LB media (which is used for bacteria as shown in Table 4.6). The mixture used is shown in Table 4.8.

Table 4.8: Chemicals in the media used for THP-1 monocyte cell growth.

Immune Cell Media	
Component	Amount
RPMI 1640	434.5 mL
10% fetal bovine serum	50 mL
1 mM sodium pyruvate	5 mL
2 mM L-Glutamin	5 mL
2 µM Mercapthoethnol	0.5 mL
1% penicilin streptomycin	5 mL

The growth and cultivation of THP-1 cells is different to that of bacterial cell cultures. THP-1 cells are stored at -193°C for a long time. After thawing, cells need to be continuously monitored and fed. The laboratory used for cell expansion is a BSL2 cell culture lab and contains an incubator at 37°C with 95% air and 5% CO_2. THP-1 are suspension cells and are thus grown within flasks with filter bottle caps on top for aeration. The cell density is always kept between 10^5 and 10^6 cells/mL and is counted using the cell counter Countess II and the Countess chamber slide. The liquid loaded on the slide is a mixture of 20 µL of cell suspension and 20 µL of trypan blue.

Media replacement and resuspension of the cells in a larger media volume typically occurs every 3-4 days depending on the measured cell culture density. The media is also a pH meter which is red/pink when fresh and yellow/orange when it needs to be changed. In order to change the media, the cell suspension is centrifuged at 150 xg for 5 min and resuspended in fresh 37°C media to a cell concentration of 10^5 cells/mL.

4.5.1 THP-1 Monocyte Stimulation

Figure 4.16 shows the THP-1 monocyte stimulation procedure. The monocyte cells are grown in suspension, and become adherent to surfaces when differentiated to macrophages.

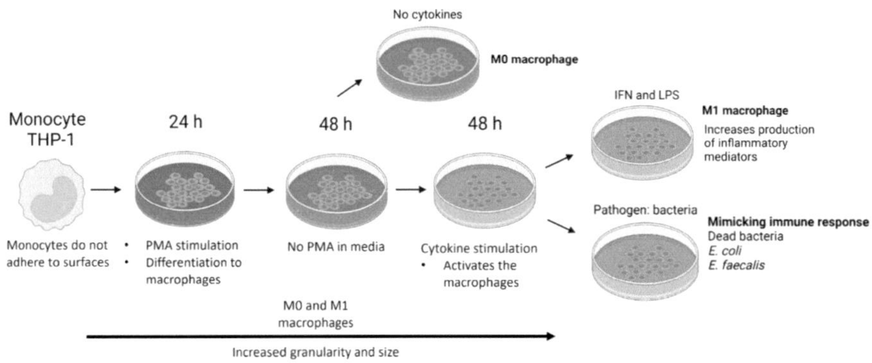

Figure 4.16: Stimulation procedure of THP-1 monocytes to macrophages. Sketch created with BioRender.com. Figure drawn based on literature [128].

1 mL of fresh media with 10^6 suspended THP-1 cells is placed in a 6 well plate and 1 mL of fresh media is added on top. 2 µL of 2 µg/mL phorbol 12-myristate 13-acetat (Sigma, 16561-209-8), PMA, is added to the well for a total concentration of 2 ng/mL in each well containing 10^6 THP-1 monocytes. After 24 h of stimulation, the cells increase granularity, become elongated and show adherence to the well bottom which indicate cell differentiation from monocytes to macrophages. The media is removed from the well and fresh media without PMA is added to the well. After 48 h, the cells can be collected for measurements of M0 macrophages. To polarize M0 macrophages to M1 macrophages, cells were treated with lipopolysaccharides (LPS) and interferon γ (IFN-γ). Media is mixed with LPS and IFN-γ for a total concentration of 1 µg/mL of LPS and 20 ng/mL. The previous media is removed and 2 mL of media mixed with LPS and IFN-γ is added to the wells. At this point, the samples can be directly removed or further stimulated in the well for 48 h before removal.

In the case of monocyte stimulation within the microfluidic chip, the cells are loaded into the microfluidic chip 24 h after PMA stimulation as indicated by the arrow drawn in Figure 4.17. The adhering cells are removed using Accutase as described in chapter 4.5.2 after PMA stimulation and pipetted into a fibronectin coated microfluidic chip (chapter 4.5.4). After 24 h, media mixed with the stimulants is added into the channel replacing the media. After 48 h, the cells are removed using Accutase for further analysis.

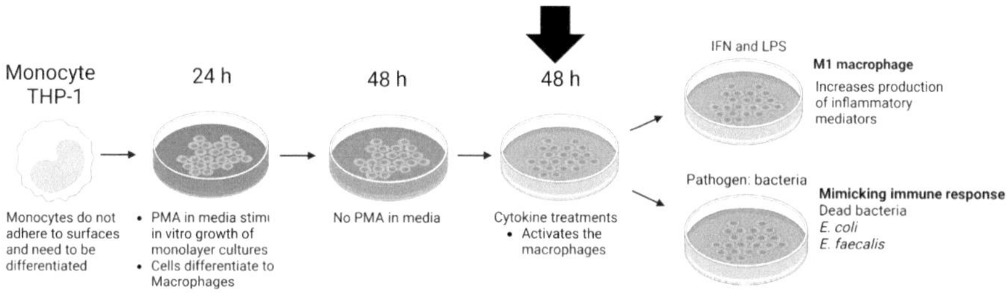

Figure 4.17: Stimulation pathways of THP-1 monocytes to macrophages in a microfluidic chip. Sketch created with BioRender.com. Figure drawn based on literature [128].

4.5.2 Macrophage Removal

Accutase (StemPro, Gibco) and sterile phosphor buffered saline (PBS) is used for the removal of adherent cells. The protocol as described in Table 4.9 is used for the collection of macrophages. Upon comparison between Accutase, Trypsin and mechanical scarping for the removal of the macrophages, Accutase has been determined to allow for the highest cell viability and amount of THP-1 cells.

Table 4.9: Steps for the removal of adherent THP-1 macrophages.

Accutase Macrophage Removal Protocol	
Step	Description
1	Remove media in the well
2	Add 1 mL of PBS
3	Remove PBS and add 1 mL of Accutase
4	Place well inside the incubator for 25 min
5	Check adherence under the microscope and increase incubator time if necessary
6	Add fresh media to the wells
7	Collect and centrifuge the cells at 150 xg for 5 min
8	Resuspend pellet in fresh media

4.5.3 Preparation for FACS

Following the stimulation pathway above, different samples are prepared for FACS analysis by staining the cells. Prior to the preparation of the plates, FACS buffer is prepared as shown in Table 4.10.

Table 4.10: Chemicals to make FACS Buffer.

FACS Buffer	
Component	Amount
PBS (pH 7.4)	500 mL
50 mM EDTA	50 mL
FBS (2%)	11 mL

The different concentrations of the fluorescent dyes is shown in Table 4.11. The concentrations can be prepared individually or all the antigen stains HLA DR and HLA ABC, and the costimulatory molecules CD8 and CD86 in one mixture.

Table 4.11: Different FACS fluorescently labelled antibodies and their concentrations in FACS buffer.

FACS Fluorescent Dye Concentration		
Dye	Fluorescence	Concentration
live/dead	510	1:200
HLA DR	PECy7	1:400
HLA ABC	APC	1:400
CD80	FITC	1:20
CD86	PE	1:100

When the different fluorescent dye concentrations and the FACS buffer are prepared, the layout for the 96 well FACS plate is made. The different wells are stained with their respective dyes. At the end of the staining, the cells are fixated using 4% paraformaldehyde (PFA) in FACS buffer. The step-by-step protocol for FACS measurements in a 96 well is shown in Table 4.12.

Table 4.12: Steps for the fluorescent staining of the FACS plate.

FACS Staining Protocol		
Step	Description	Time
1	Add 125 µL of each sample to be measured into the wells	5 min
2	Centrifuge well at 500 xg for 5 min and discard 90 µL	5 min
3	Pipette 100 µL of fluorescent antigen dilution into the wells which are to be stained - either with the mixture of all the antigens or individual antigens. Fill the wells which are not to be stained with FACS buffer. Mix each well thoroughly	10 min
4	Keep the plate at 4°C for 30 min in the dark	30 min
5	Add 100 µL of FACS buffer and resuspend	5 min
6	Centrifuge the plate at 500 xg for 5 min	5 min
7	Remove 200 µL and leave pellet	5 min
8	Pipette 100 µL of fluorescent live/dead marker dilution into the wells which are to be stained. Fill the wells which are not to be stained with FACS buffer. Mix each well thoroughly	10 min
9	Keep the plate at room temperature for 30 min in the dark	30 min
10	Add 100 µL of FACS buffer and resuspend	5 min
11	Centrifuge the plate at 500 xg for 5 min	5 min
12	Remove 200 µL and leave pellet	5 min
13	Add 100 µL of 4% PFA in FACS buffer and resuspend	5 min
14	Keep the plate at room temperature for 10 min in the dark	10 min
15	Add 100 µL of FACS buffer and resuspend	5 min
16	Centrifuge the plate at 500 xg for 5 min	5 min
17	Remove 200 µL and leave pellet	5 min
18	Add 100 µL of FACS buffer and resuspend	5 min
19	Keep at 4°C in a dark place	Until use

4.5.4 Microfluidic Chip Preparation

The microfluidic chips are coated with fibronectin to aid the adhesion of the immune cells.

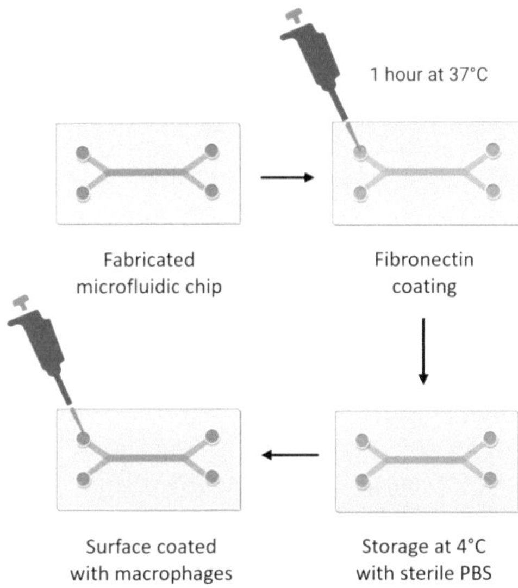

Figure 4.18: Preparation steps for the PDMS microfluidic chip for the adhesion of THP-1 differentiated macrophages. Sketch created with BioRender.com.

The chips are coated with fibronectin within an hour of oxygen plasma ashing the slide and the PDMS together. 200 µL pipette tips are inserted into the outlets and 120 µL of 100 µg/mL fibronectin is filled into the chip. The chip is subsequentely placed in the incubator at 37°C for 30 min. Following the coating step, the chip is washed with sterile PBS 3 times. During the washing steps it is important to make sure that no air bubbles are introduced into the microfluidic channel. For storage, the chip is filled with sterile PBS and the inlets and outlets have 200 µL maximum volume pipette tips also filled with sterile PBS. The pipette tips are sealed with Parafilm to ensure that the PBS does not evaporate and dry out the fibronectin coated channels. The chip is stored at 4°C for a maximum of one week. Check for any air bubbles every 3-4 days.

5 Experimental Determination of Heat Originating from Cells

To experimentally determine the *in vitro* heat flow originating from bacterial cells on an infected implant, a calorimetric microfluidic chip is used. The microfluidic chip experiments can be classified in two general different layouts: one channel level microfluidic chip as described in chapter 4.1.1, 5.1.2, and 5.3 and double channel microfluidic chip as described in chapter 4.1.2 and 5.2. In the one channel level microfluidic chip, both chemical exothermic experiments and biological experiments measuring bacterial growth are shown. The exothermic chemical calibration of the microfluidic chip was done using a base catalyzed hydrolysis of methyl paraben, which is a first order exothermic chemical reaction often used for thermal calibration of the microcalorimetric chip [127] as also described in chapter 5.1.2. This experiment is used in combination with a microcalorimetric experiment (chapter 5.1.1) to determine the heat transfer efficiency of the microfluidic chip as described in chapter 5.1.2.2. This is defined as the percentage ratio of heat flow measured by the sensor to the heat flow originating form the sample. The growth of *E. coli* is measured in the microfluidic chip through the thermal power originating from cell metabolism in addition to optical density measurements counting the amount of present cells as described in chapter 5.1.2.3. Further, a double channel microfluidic chip is presented, with the purpose of moving a step closer to the biological model system of an infected vascular graft by emulating the thermal environment of the implant with an additional convective heat transfer element. In this system, the chemical exothermic reaction is applied detailed in chapter 5.2. The measurement of bacterial growth in the microfluidic channel is repeated in a thermally not stabilized environment to further approach the *in vivo* thermally fluctuating system of the thorax as described in chapter 5.3. A color version of this chapter is available at https://doi.org/10.3929/ethz-b-000703739.

As discussed in the state of the art, bacterial growth is not the only contributing heat flow source at a growing pathogenic infection on an implant. Immune cells increase their metabolism in contact with pathogens and therefore also increase their heat flow. To investigate this, different types of stimulated immune cells are investigated by fluorescence activated cell sorting (FACS), as shown in chapter 5.4.1, and the heat flow originating from these cells is measured using microcalorimetry as shown in chapter 5.4.2. These experiments allow for further insight in the feasibility of the early detection of infection on implant surfaces through heat flux measurements.

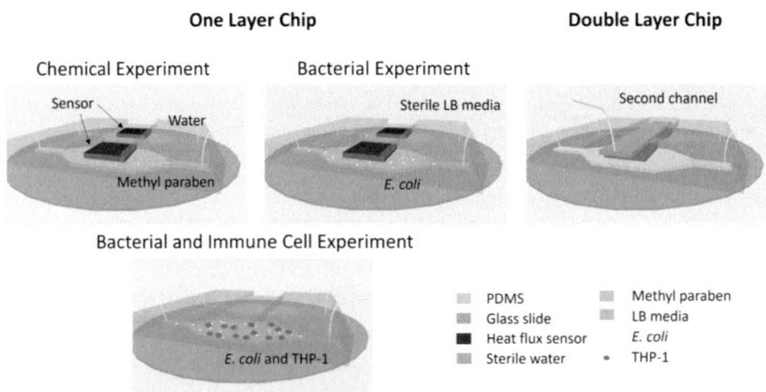

Figure 5.1: Overview over the different microfluidic experimental configurations presented in this chapter. The one layer chip is used for both a chemical exothermic experiment, a bacterial experiment and a co-culture of a bacteria and immune cells. The double layer chip is used as an *in vitro* system for the aortic vascular graft. Figure based on literature published under the Open Access [122, 123] under CC BY license (Creative Commons Attribution 4.0 International license).

5.1 Microfluidic Chip Heat Transfer Efficiency Determination

This chapter is based on previously published work [122]. To determine the heat transfer efficiency of the microfluidic chip, the measurement of the exothermic chemical reaction of methyl paraben with sodium hydroxide is used. As a reference measurement, the total heat flow originating from the chemical exothermic reaction is measured using a microcalorimetric setup. Comparing the total measured heat flow originating from the exothermic chemical reaction to the heat flow measured in the microcalorimetric chip with the same exothermic chemical reaction, the heat transfer efficiency of the microfluidic chip can be determined. The heat transfer efficiency dictates the ratio of the heat flow measured by the sensor to the total heat flow originating from the channel. This is a characterization experiment for the determination of the total heat flow originating from the microfluidic channel for biological measurements. In the following chapter 5.1.1, the total heat originating from by 20 µL, 100 µL, and 200 µL of the exothermic base catalyzed hydrolysis of methyl paraben is determined through microcalorimetry. Subsequently, in chapter 5.1.2.1 the heat transfer efficiency is experimentally determined in both channels of the microfluidic chip.

5.1.1 Microcalorimetric Reference Measurement

The thermal calibration experiment is set up according to the description by O'Neill et al. [127] using the first order reaction of NaOH catalyzed hydrolysis of methyl paraben at 37°C which is published in literature [122]. The detailed experimental procedure is presented in chapter 4.3.1. A microcalorimetric experiment (commercial microcalorimeter TAMII nano (Waters/TA, Delaware, USA)) for three different volumes, 20 µL, 100 µL, and 200 µL as shown in Figure 5.2 is performed. The ratio between the volumes is the same as the ratio between the heat flow measured by these volumes at each time point considered. The y-axis intercept at t = 0 s is the total heat flow at the beginning of the experiment for the given volume, and is used as a reference point to compare the microcalorimetric system with the experiment in the microfluidic chip. The microcalorimetric experiment serves as the reference heat flow per volume and the slope of heat flow decrease is k [1/s]. According to literature, the slope, k, changes with the temperature at which the experiment is performed [127]. At 37°C, k is -7.75×10^{-4} 1/s and at 24°C k is -3.4×10^{-4} 1/s [127]. The extracted slopes of Figure 5.2 are shown in Table 5.1.

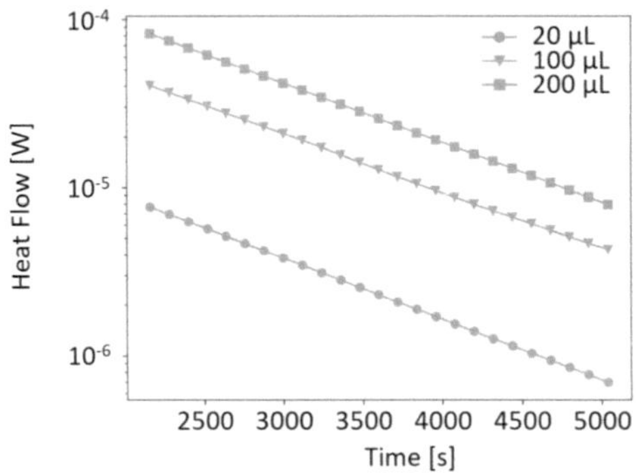

Figure 5.2: Semi-logarithmic data of the microcalorimetric experiment of different volumes of samples (20 µL, 100 µL, and 200 µL) of the base catalyzed hydrolysis of methyl paraben at 37°C using a commercial microcalorimeter TAMII nano (Waters/TA Delaware, USA). t=0 s is the time at which NaOH is mixed with methyl paraben. The first 2000 s were used for the thermal stabilization of the microcalorimeter. Open Access figure adapted from [122] published under open access under CC BY license (Creative Commons Attribution 4.0 International license).

The data in Figure 5.2 is fitted with the general function:

$$ln(\dot{Q}) = ln(\dot{Q}_0) + kt \tag{5.1}$$

Where $\dot{Q}$ is the measured heat flow by the microcalorimeter, $\dot{Q}_0$ is the heat flow at the y axis intercept, k is the slope of the heat flow increase, and t is the time with t=0 s as the time at which the NaOH is mixed with methyl paraben. Table 5.1 shows the extracted slope of the different initial sample volumes from Figure 5.2.

Table 5.1: Extracted parameters of the methyl paraben reference measurement in the microcalorimeter.

Extracted Parameters		
System	Intercept [W]	Slope [1/s]
20 µL	4.63×10^{-5}	-8.32×10^{-4}
R^2	1.0000	
100 µL	2.22×10^{-4}	-7.86×10^{-4}
R^2	0.9999	
200 µL	4.73×10^{-4}	-8.10×10^{-4}
R^2	1.0000	

5.1.2 One Channel Level Microfluidic Chip

The following section is based on a published paper [122]. Chapter 4.1.1 describes the experimental setup developed for measuring the exothermic chemical calibration experiment and measurement of the thermal activity of bacteria. The heat flux measurement is based on a differential microfluidic calorimetry chip with two integrated heat flux sensors (gSKIN XP 26 C from greenTEG, datasheet in Appendix chapter C.1). The microfluidic chip is fabricated as described in chapter 4.1.1 with channel length of 12 mm and channel height of 320 µm as shown in Figure 4.4 a). The heat flux sensor is placed above the channel with a 150 µm spacing of PDMS in between.

The working principle of the calorimetric microfluidic chip is similar to a flow microcalorimeter [127]. Fluid containing the sample is transported from the containers through tubing to the channels using a peristaltic pump. Using two heat flux sensors (one without the sample and one with) enables differential compensation of the heat flux, which reduces external thermal common-mode fluctuations. For this system, a PMMA box and a copper piece is used for further local stabilization of the environmental thermal fluctuations, especially of importance since opening the oven door to add the methyl paraben with NaOH creates an additional temperature change.

5.1.2.1 One Channel Level Microfluidic Chip Chemical Calibration

As described in the experimental setup chapter 4.2.1, one of the channels (underneath sensor 1) continuously is filled with DI water and the other channel (underneath sensor 2) is at some point filled with the chemical exothermic reaction. This experiment is then repeated with the channels filled with opposite contents after a cleaning step as shown in Figure 4.7.

At the beginning of the experiment, both channels are filled with DI water for the calibration step as shown in Figure 5.3 a) and 5.4 a). The purpose of the calibration is the determination of the baseline offset between the two sensors which can be corrected (correction in the calibration phase shown in Figure 5.3 c) and 5.4 c)). The approach is to correct one of the sensor outputs and subsequently subtracting the difference between the two values yielding a differentially compensated heat flux signal, which in the calibration phase is around 0. Given two heat flux signals with q_{c1} (with the subscript $_1$ or $_2$ representing the different sensors) representing the channel which is later to be filled with the exothermic chemical reaction (purple color) and q_{c2} representing the calibration channel (blue color). When one of the channels is filled with the exothermic chemical reaction, the other channel still contains DI water. The purpose of the channel containing DI water is to measure the thermal fluctuations of the surrounding environment. By applying the correction, a differentially compensated heat flux signal solely representing the heat flow output of the exothermic chemical reaction is extracted (completely independent of background thermal fluctuation). Two experiments are performed, where the contents of the two channels are interchanged (Figure 5.3 vs. Figure 5.4) as illustrated in Figure 4.7.

In the calibration phase where both channels are filled with water, the difference should therefore be around 0. The subscript $_c$ indicates the heat flux data from the calibration phase. The corrected heat flux signal q^*_{c2} is expressed as:

$$q^*_{c2} = L_c(q_{c2} - \bar{q_{c2}}) + \bar{q_{c2}} + \mid \bar{q_{c1}} - \bar{q_{c2}} \mid \tag{5.2}$$

$$q^*_{c2} = L_c(q_{c2} - \bar{q_{c2}}) + \bar{q_{c1}} \tag{5.3}$$

Where L_c as the ratio between the standard deviation of the two heat flux signals, σ_{c1}/σ_{c2}. $\bar{q_{c1}}$ and $\bar{q_{c2}}$ are the averaged values of the heat flux signal over 10 data-points. The differentially compensated heat flux signal of the calibration phase q^*_{c2} is shown in Figure 5.3 c) and 5.4 c) in blue.

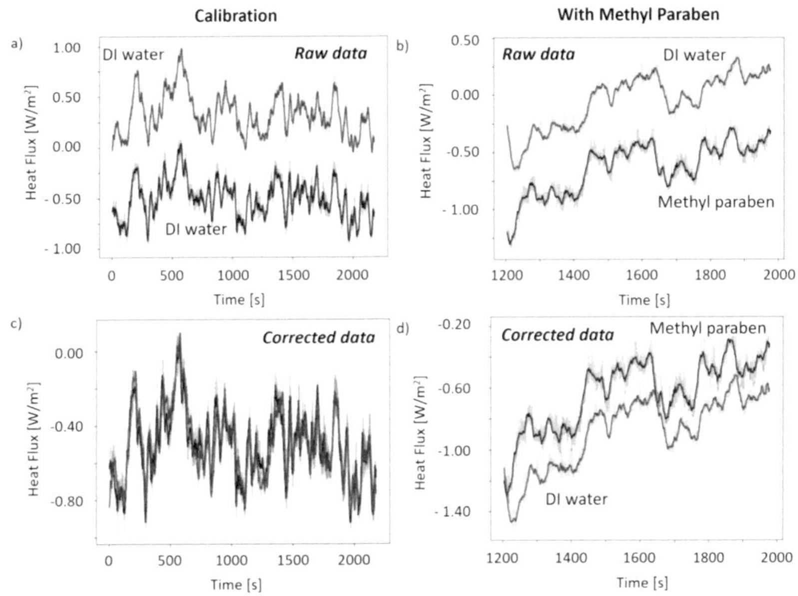

Figure 5.3: Methyl paraben experiment with microfluidic calorimetric chip with methyl paraben in sensor 1. a) and c) show the heat flux signals for the two sensors in the calibration phase where both channels are filled with DI water. The blue signal is the control channel and the purple signal is the channel in which the exothermic chemical reaction is added. b) and d) show the data upon addition of the exothermic chemical in the channel. c) and d) show the data after the differential correction of the sensor signal. Open Access figure adapted from [122] published under open access under CC BY license (Creative Commons Attribution 4.0 International license).

To extract the heat flux by the chemical exothermic reaction, the difference between q_1 and q_2^* is calculated. The subscript $_m$ indicates the data of the two heat flux sensors in the methyl paraben experiment phase:

$$q_m = q_{m1} - q_{m2}^* \tag{5.4}$$

with

$$q_{m2}^* = L_c(q_{m2} - \bar{q_{c2}}) + \bar{q_{c1}} \tag{5.5}$$

Where q_{m1} is the heat flux signal in the channel filled with methyl paraben and q_{m2} is the heat flux signal by sensor above the microfluidic channel filled with DI water. The extracted corrected difference between the two

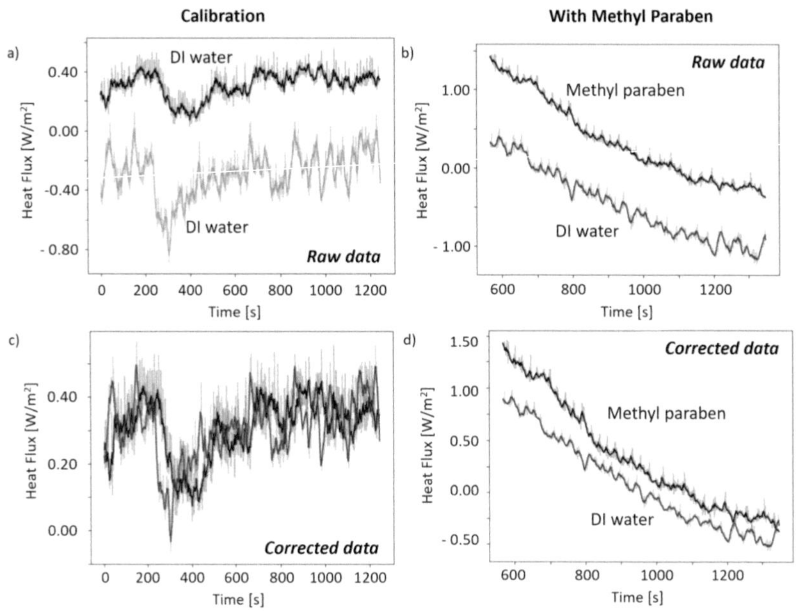

Figure 5.4: Methyl paraben experiment with microfluidic calorimetric chip in the opposite channel as is shown in 5.3 with sensor 2 filled with methyl paraben. a) and c) show the heat flux signals for the two sensors in the calibration phase where both channels are filled with DI water. The blue signal is the control channel and the purple signal is the channel in which the exothermic chemical reaction is added. b) and d) show the data upon addition of the exothermic chemical in the channel. c) and d) show the data after the differential correction of the sensor signal. Open Access figure adapted from [122] published under open access under CC BY license (Creative Commons Attribution 4.0 International license).

heat flux sensors, q^*_{m2}, is also referred to as the differentially compensated heat flux signal. Figure 5.5 a) and b) show the differentially compensated heat flux signals measured from the exothermic chemical reaction in both experiments as shown in Figure 4.7.

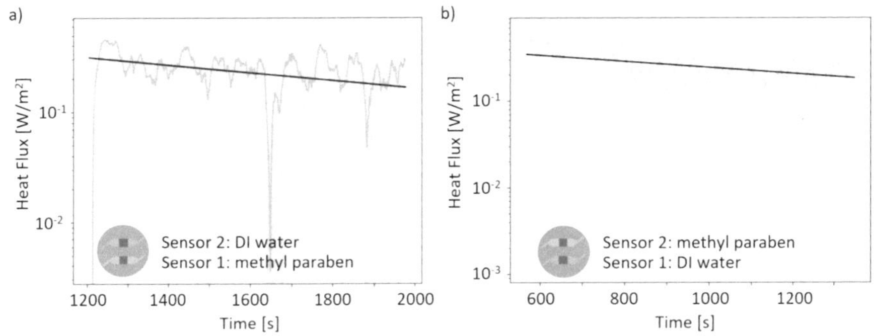

Figure 5.5: Differentially compensated heat flux signal measured from the exothermic chemical reaction of methyl paraben. Open Access figure adapted from [122] published under open access under CC BY license (Creative Commons Attribution 4.0 International license).

Both slopes in Figure 5.5 are fit with -8.09×10^{-4} 1/s which is close to the literature value of -7.75×10^{-4} 1/s [127] and similar to the measured slopes shown in Table 5.1 which range between -7.86×10^{-4} 1/s and -8.32×10^{-4} 1/s.

5.1.2.2 Comparison Between Microcalorimetric Experiments and Microfluidic Exothermic Chip Experiments

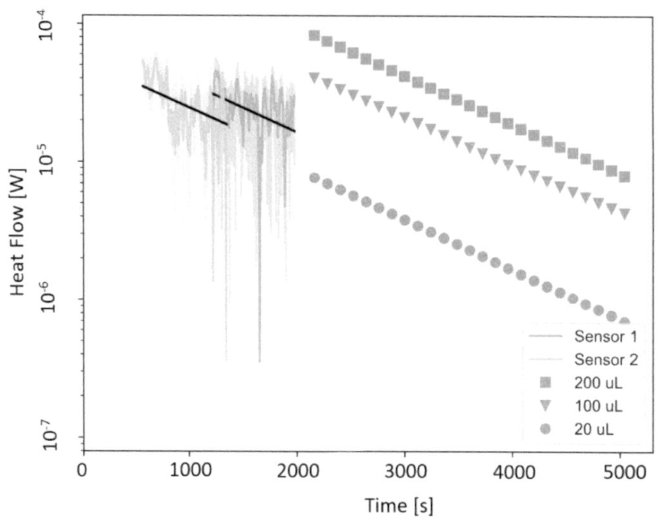

Figure 5.6: Comparison between microcalorimetric and microfluidic calorimetric experiments. The data is from Figure 5.2 a) and Figure 5.5 a) and b). The measurement is shifted to have t = 0 s being the time at which NaOH is added to methyl paraben. Open Access figure adapted from [122] published under open access under CC BY license (Creative Commons Attribution 4.0 International license).

Figure 5.6 shows the reference data measured by the microcalorimeter in Figure 5.2 and the data measured in the microfluidic chip in Figure 5.5. The heat flux is converted to heat flow as follows $\dot{Q} = A_{sensor} q_m$. t = 0 s is the point at which NaOH is mixed with methyl paraben in both experiments. To determine the heat transfer fraction of the two different microcalorimetric channels, the heat flow measured in the microfluidic chip of 46 µL is compared to the total heat flow originating from the microcalorimetric samples of 20 µL, 100 µL, and 200 µL.

The heat transfer efficiency is defined as:

$$X_{mp} = \dot{Q}_{m0}/\dot{Q}_0^{\mu} \tag{5.6}$$

With $\dot{Q}_{m0}$ as the y-axis intercept of the microfluidic channel and $\dot{Q}_0^{\mu}$ as the y-axis intercept of the reference microcalorimetric measurement scaled to the volume of the channel of 46 µL. The subscript $_m$ indicates the microfluidic methyl paraben experiment, the subscript $_0$ indicates the y-axis intercept, and the superscript $^{\mu}$ indicates the microcalorimetric measurement.

The heat transfer efficiency, X_{mp}, also relates the total exothermic heat flow originating from the channel to the actual heat flow measured by the heat flux sensor (and will be expressed as a percentage). Thus, it can be used as a conversion parameter for the determination of the total heat flow originating from the channel for further heat flux measurements of the microfluidic chip.

For the microfluidic channel under sensor 1 the determined heat transfer efficiency is 77% ± 8% and for the microfluidic channel under sensor 2 it is 52% ± 9% (error propagation discussed in the Appendix in chapter E.3). The varying heat transfer efficiency could be due to fabrication differences between the two channels in addition to the possible temporary creation of air bubbles in the channel. The determined values are general in agreement with the determined heat transfer efficiency of 73% of the one layer microfluidic system as described in chapter 3.4.1.1.

5.1.2.3 One Channel Level Microfluidic Chip Bacterial Experiment

In the bacterial experiments, *E. coli* MG1655 (shown in Figure 5.7) is used as the model biological system. In the microfluidic system, the growing bacterial population in the channel mimics a growing infection on an implant surface. Finally, the heat flux limit of detection of the bacterial population is compared to possible early detection of infection on an implant surface.

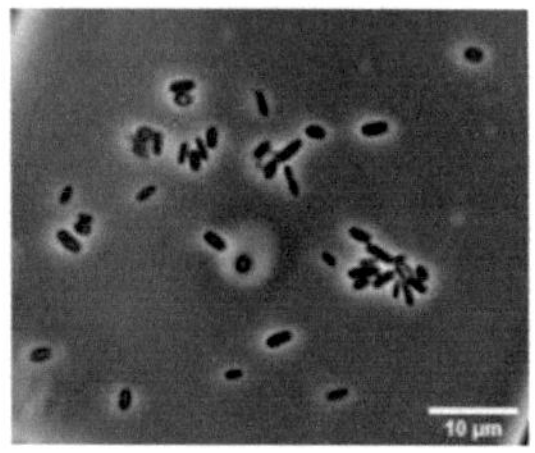

Figure 5.7: *E. coli* bacteria strain MG 1655 used in the bacterial experiments.

Similar to the methyl paraben experiment, the bacterial experiment measures bacterial growth in both channels (sensor 1 channel and sensor 2 channel). There is minimal influence of the anaerobic growth conditions within the tubing. To show this, a comparison between the aerated bacteria in the flask and bacteria passed through tubing and the microfluidic chip is shown in the Appendix in chapter E.1. Initially, the experiment has a calibration phase where both channels are filled with sterile LB media. Subsequently bacteria is added to one of the channels. In between the two experiments both channels were sterilized with 70% ethanol in sterilized water as shown in Figure 4.10.

In the bacterial experiment, a heat flux signal, q, of the bacterial growth is extracted by applying the same differential compensation method on the heat flux sensors as shown above for the chemical calibration and control/compensation (LB) heat flux data. In Figure 5.8 and Figure 5.9 a), b), c), and d) the calibration phase is shown with both channels containing media without bacteria. However, the contents of the channels is flipped in these two experiments as shown in Figure 4.10. In all figures, the raw data is shown in the lighter color, and a 200 point moving-average is shown in the darker color. In this experiment, the differential compensation method is as follows:

$$q_{c2}^{*} = L_c(q_{c2} - \bar{q_{c2}}) + \bar{q_{c2}} + \mid \bar{q_{c1}} - \bar{q_{c2}} \mid \tag{5.7}$$

This is the same as Equation 5.3.

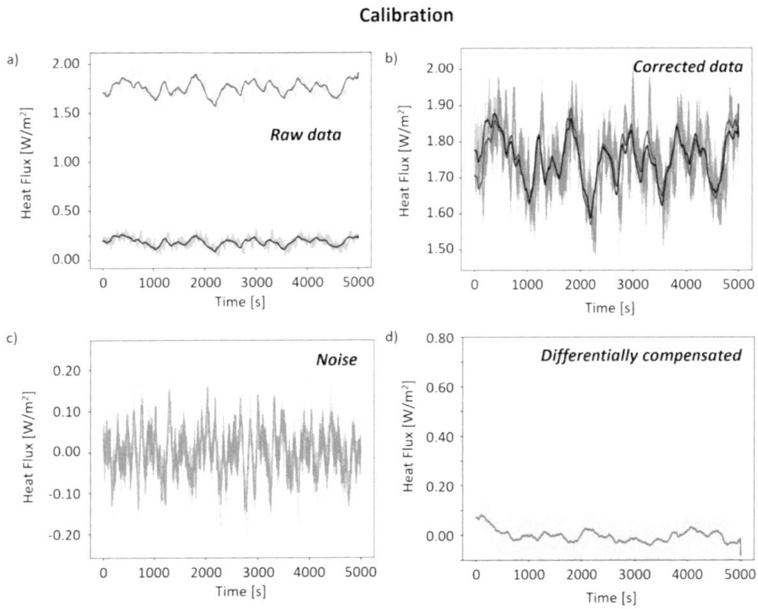

Figure 5.8: Bacterial experiment with microfluidic calorimetric chip with LB media in both channels under sensor 1 and sensor 2. a) shows the heat flux signals in both channels for sensor 1 and sensor 2, with sensor 1 indicating the channel in which bacteria will be added in a later step. b) shows the applied differential compensation on the signal of sensor 2. c) shows the noise left upon correction of the heat flux signals. d) shows the differentially compensated heat flux in the calibration phase which is the difference between the two sensors in b). Open Access figure adapted from [122] published under open access under CC BY license (Creative Commons Attribution 4.0 International license).

The heat flux measured upon the addition of bacteria is shown in Figure 5.10 (sensor 1 contains bacteria, sensor 2 with only media). After a sterilization step, the experiment was repeated, reversing the function of the sensing and control sensor with data shown in Figure 5.11 (sensor 2 contains bacteria sensor 1 with only media). The corrected heat flux signal is shown in b) in both figures. In Figure 5.10 d) and Figure 5.11 d) the differential heat flux measured by the bacterial growth as defined by:

$$q_{b2}^{*} = L_c(q_{b2} - q_{\bar{b}2}) + \bar{b_{c2}} + \mid q_{\bar{c}1} - q_{\bar{c}2} \mid \quad (5.8)$$

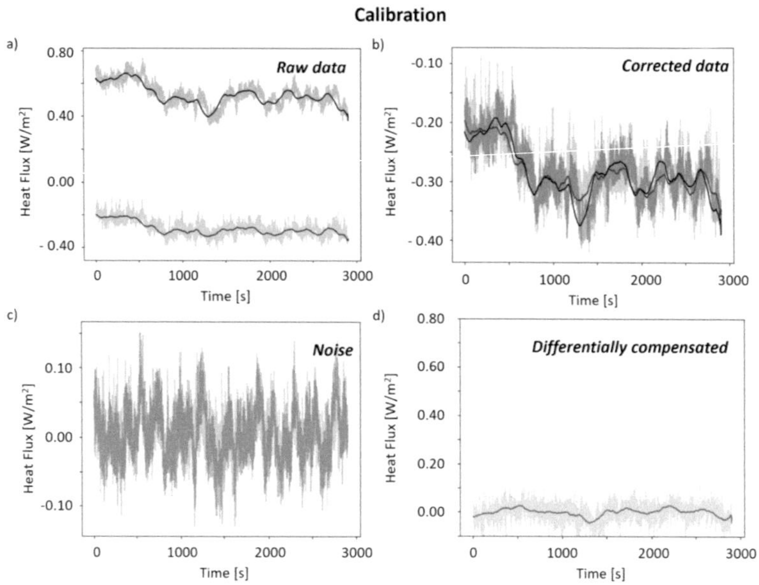

Figure 5.9: Bacterial experiment with microfluidic calorimetric chip with LB media in both channels under sensor 1 and sensor 2. This calibration phase is after a sterilization step of 70% EtOH. a) shows the heat flux signals in both channels for sensor 1 and sensor 2, with sensor 2 indicating the channel in which bacteria will be added in a later step. b) shows the applied differential compensation on the signal of sensor 2. c) shows the noise left upon correction of the heat flux signals. d) shows the differentially compensated heat flux in the calibration phase which is the difference between the two sensors in b). Open Access figure adapted from [122] published under open access under CC BY license (Creative Commons Attribution 4.0 International license).

The differentially compensated heat flux in the calibration experiment are defined as $q_b = q_{b1} - q_{b2}^*$.

The data is averaged over 200 data points with a sampling frequency of 0.7 Hz. With this averaging, the noise in the differential heat flux data is significantly reduced from σ_q=34 mW/m^2 to σ_q=15 mW/m^2. The averaging time is still significantly smaller than the expected doubling time of the bacterial population (showed in Table 5.2), $\tau_{averaging} \ll \tau_{bacteria}$.

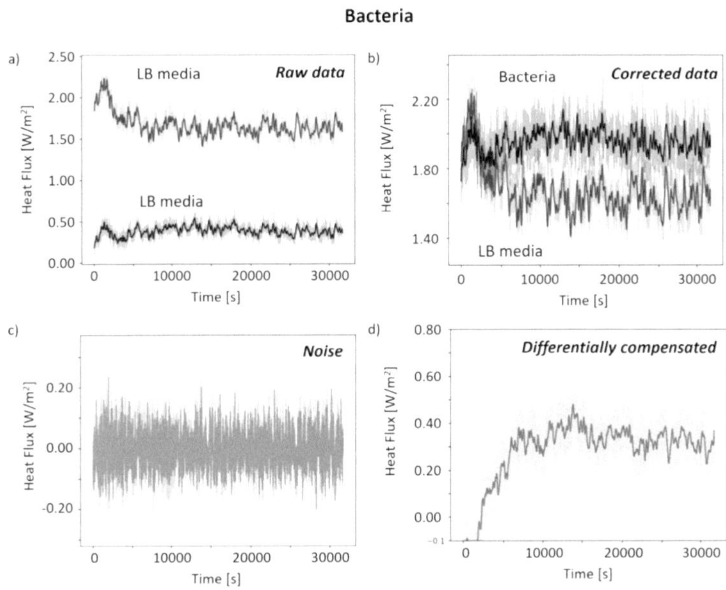

Figure 5.10: Bacterial experiment with microfluidic calorimetric chip with bacteria in the channel under sensor 1 and LB media in the channel under sensor 2. a) shows the heat flux signals in both channels for sensor 1 and sensor 2, with sensor 1 indicating the channel in which bacteria is added at t = 0 s. b) shows the applied differential compensation on the signal of sensor 2. c) shows the noise left upon correction of the heat flux signals. d) shows the differentially compensated heat flux measuring the thermal power originating from bacterial growth. Open Access figure adapted from [122] published under open access under CC BY license (Creative Commons Attribution 4.0 International license).

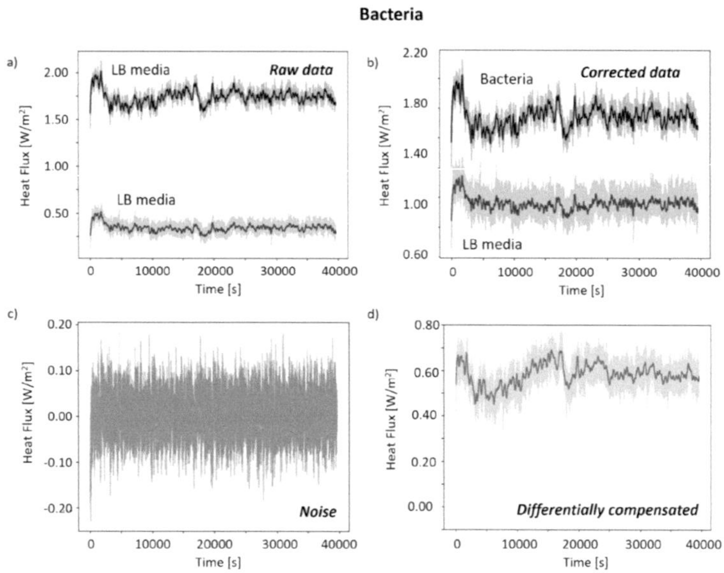

Figure 5.11: Bacterial experiment with microfluidic calorimetric chip with bacteria in the channel under sensor 2 and LB media in the channel under sensor 1. a) shows the heat flux signals in both channels for sensor 1 and sensor 2, with sensor 1 indicating the channel in which bacteria is added at t = 0 s. b) shows the applied differential compensation on the signal of sensor 2. c) shows the noise left upon correction of the heat flux signals. d) shows the differentially compensated heat flux measuring the thermal power originating from bacterial growth. Open Access figure adapted from [122] published under open access under CC BY license (Creative Commons Attribution 4.0 International license).

To add bacteria, the incubator door must be temporarily opened causing thermal fluctuations at the beginning of the experiment. A PMMA box and a copper piece are used on top of the microfluidic chip to compensate for the thereby induced thermal fluctuation. In these two experiments, the heat flux signal of sensor 2 (blue in Figure 5.10 a) and purple in Figure 5.11 a)) shows a larger response to environmental thermal perturbation. This thermal perturbation is not completely compensated by the differential compensation scheme resulting in the initial $\approx$ 4000 s of the data shown in Figure 5.10 d) and 5.11 d) are affected. Due to the nature of the differential compensation, the perturbation causes an initial additional decrease in the differentially compensated heat flux signal for Figure 5.10 d) and an initial additional increase in the differentially compensated heat flux signal for Figure 5.11 d).

5.1.2.4 Extraction of Bacterial Thermal Values

The OD measurement of bacteria, collected using a piece of tubing from the outlet of the microfluidic channel fed through a slit in the incubator to the outside of the incubator in parallel to the bacterial heat flux measurement allows the determination of total bacterial population, N_t, as a function of time as also shown in Figure 5.12. Although the growth conditions were identical in both channels, the growth of *E. coli* is slightly different in the two experiments, which could be due to differentiating initial bacterial concentrations. N_t is the optical density measured during the exponential growth phase, OD_{ratio} is the conversion from optical density to bacteria/mL as described in the Appendix in chapter E.2. In the exponential growth phase, as defined by the optical density measurements, the data can be fitted by the function:

$$N_t = N_0 e^{\mu_{OD} t} \tag{5.9}$$

$$ln(N_t) = ln(N_0) + \mu_{OD} t \tag{5.10}$$

With N_t as the bacterial population at time, t, N_0 as the bacterial population at time, $t = 0$, and μ_{OD} as the slope/time constant of the exponential growth phase.

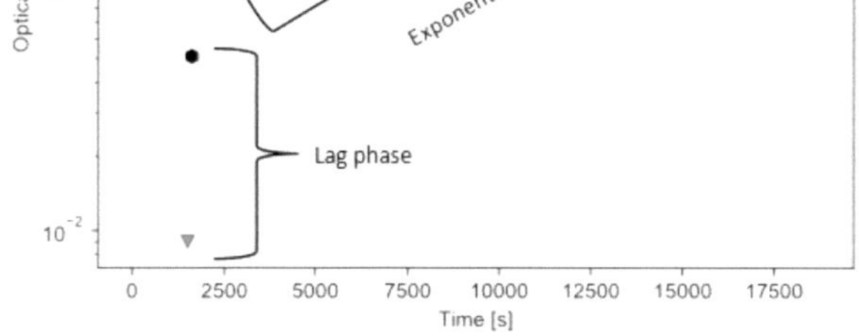

Figure 5.12: Optical density measurement as shown for Figure 5.10 (Sensor 1) and 5.11 (Sensor 2). The samples were placed on ice directly upon collection and measured on an OD meter.

The doubling time of the bacteria in the exponential growth curve is be defined by:

$$G = \frac{t_2 - t_1}{generations} = \frac{t_2 - t_1}{3.3log(\frac{N_2}{N_1})} \tag{5.11}$$

With G as the doubling time, t_2 and t_1 as two different time points in the exponential growth phase with corresponding optical density measured bacterial populations N_2 and N_1.

In the same defined exponential growth region determined by the OD measurement, the heat flow $\dot{Q} = qA_{sensor}$ signal is also fit to an exponential function:

$$\dot{Q} = \dot{Q}_0 e^{\mu_Q t} \tag{5.12}$$

$$ln(\dot{Q}) = ln(\dot{Q}_0) + \mu_Q t \tag{5.13}$$

The exponential regions are shown by the drawn lines in Figure 5.13 comparing the optical density measurements with the heat flux measurements. This is the same data as plotted in Figure 5.10 d) and Figure 5.11 d) only difference is that it is in a semi-logarithmic plot with different y-axis limits.

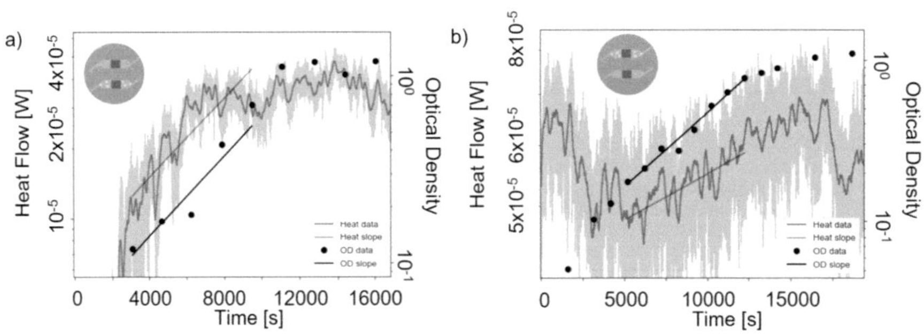

Figure 5.13: Semi-log plots of the differentially compensated heat flow of both channels as shown in 5.10 d) and 5.11 d) with an indication of the exponential growth region of both the heat flow measurement and the Optical Density measurement. Open Access figure adapted from [122] published under open access under CC BY license (Creative Commons Attribution 4.0 International license).

Table 5.2 compares the different values in the two experiments as shown in Figure 5.13. The first column indicates the sensor under which the bacteria was placed and both experiments were conducted subsequently following a sterilization step using ethanol. The two channels respond differently to the thermal stabilization step after the incubator door was opened and bacteria was added, as was discussed above. This difference in thermal stabilization is one of the reasons for the measured difference in the heat flow slope of the two experiments. The slope of the exponential growth measured through the optical density is similar in both channels. When further comparing the optical density and the heat flux values one can extract the single thermal power of an *E. coli* bacterium as already shown in the last column of Table 5.2.

Table 5.2: The calculated doubling time and the slope of the exponential growth.

System	Doubling time G	Slope μ_{OD} [1/s]	Slope μ_Q [1/s]	Heat transfer efficiency X_{mp} [%]	$\dot{Q}_{bacterium}$ [pW]
Sensor 1	41.65 min	2.81×10^{-4}	1.97×10^{-4}	77 ± 8	1.3-4.5
Sensor 2	56.08 min	2.18×10^{-4}	2.79×10^{-5}	52 ± 9	1.9-4.0

Figure 5.14 shows a schematic overview over how the single bacterium thermal power is calculated using the measured heat flux q, optical density values N_t, and the heat transfer efficiency X_{mp}.

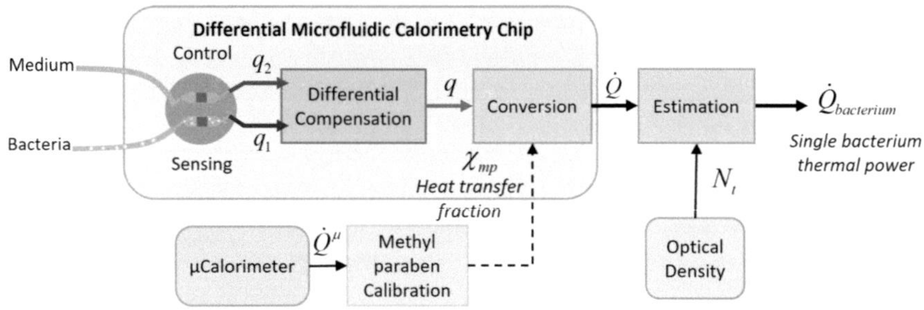

Figure 5.14: Heat flow of both channels as shown in Figure 5.13in the exponential growth region. The extracted single bacterium thermal power in both experiments. Open Access figure adapted from [122] published under open access under CC BY license (Creative Commons Attribution 4.0 International license).

The measured heat flux q [W/m^2] is differentially compensated and converted to heat flow $\dot{Q}_{bacterium}$ [W] by $q_{total} = \dot{Q}/(X_{mp}A_{sensor})$. The total heat flow in the channel is determined by using the heat transfer efficiency determined in the microfluidic chip X_{mp}. The optical density measurement of N_t is converted to the bacterial population in the channel by conversion using the OD_{ratio} (as described in the Appendix in chapter E.2.) and the volume of the channel $V_{channel}$ (46 µL). Thus, the equation for the determination of the thermal power of a single bacterium is determined as:

$$\dot{Q}_{bacterium} = \frac{\dot{Q}/X_{mp}}{N_t OD_{ratio} V_{channel}} \tag{5.14}$$

Applying this formula, the single bacterium thermal power is extracted and the values are shown in Figure 5.15. In literature, most cases show that the thermal power of a single bacterium stays constant throughout the exponential growth phase. In Figure 5.15, the extracted single bacterium thermal power decreases over time. The decrease indicate a transition from the exponential growth phase to the stationary phase which corresponds to a decrease in the metabolism of the bacteria. Furthermore, the influence of a switch in nutrient sources has shown to have an effect on the bacterial growth in LB media [129]. Further, Alklint et al. [75] has shown a concave curve for the thermal power over time for bacterial growth also indicating a non-constant single bacterial thermal power.

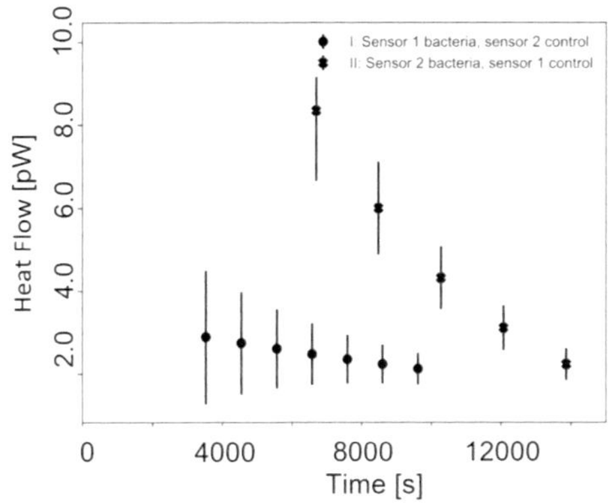

Figure 5.15: Thermal output of both channels as shown in 5.10 and 5.11 in the exponential growth region. The extracted single bacterium thermal power in both experiments is shown. Open Access figure adapted from [122] published under open access under CC BY license (Creative Commons Attribution 4.0 International license).

The determined single bacterium thermal power ranges between 1.3 pW and 4.5 pW which covers the range of values previously determined for *E. coli* bacteria of 3.5 pW [130]. The error propagation for the single bacterium thermal power is described in the Appendix in chapter E.3.

Further, with measurements of both optical density and heat flux (with a standard deviation of σ_q=15 mW/m^2) the sensitivity and the limit of detection are determined. The sensitivity in terms of thermal power per optical density of the system is determined as 6.8×10^{-14} W/bacteria (0.73 W/m^2/OD) and the limit of detection is determined as 2×10^7 bacteria/mL (0.021 OD) (as described in the Appendix in chapter E.5). This corresponds to 9.2×10^5 bacteria in the microfluidic chip, which corresponds to a limit of detection in the terms of thermal power density of 1707 W/m^3 in the 46 µL microfluidic channel (assuming a thermal power of a single bacterium of 3.5 pW [130]). Commercial microcalorimeters are able to achieve a limit of detection of 10^4 active microbial cells [72]. An extensive overview over the extracted performance parameters of the microfluidic chip can be found in the Appendix in chapter E.4. Table 5.3 shows a comparison between the presented microfluidic chip and previous systems in literature including conventional industrial calorimeters.

Table 5.3: The table is adapted from Braissant et al. [131] and is a good comparison as to the parameters of the system at hand. The extracted value is derived in the Appendix and shown in Table E.3.

Microcalorimeter type	**Volume [mL]**	**Detection limit [µW]**	**Specific sensitivity [mW×L^{-1}]**	**Ref.**
		Chip Calorimeters		
Johannessen et al.	7.2×10^{-7}	1.3×10^{-2}	1.8×10^{4}	[90]
Zhang et al.	1.5×10^{-5}	5×10^{-2}	3.3×10^{3}	[132]
Torres et al.	5.0×10^{-4}	1×10^{-1}	2×10^{2}	[88]
Lerchner et al.	5.0×10^{-3}	5×10^{-2}	1.0×10^{1}	[133]
Higuera-Guisset et al.	6.0×10^{-1}	1×10^{-1}	1.7×10^{-1}	[130]
		Conventional Calorimeters		
Calmetrix I-Cal 800	1.25×10^{2}	2×10^{1}	1.6×10^{-1}	
Waters/TA TAM	2×10^{1}	2.5	1.3×10^{-1}	
THT µMC	1.5-4.0	2×10^{-1}	1.3×10^{-1}-5.0×10^{-2}	
Waters/TA TAM48	1.0-4.0	2×10^{-1}	2.0×10^{-1}-5.0×10^{-2}	
Waters/TA TAM III	1.0-4.0	2×10^{-2}	2.0×10^{-2}-5.0×10^{-3}	
Setaram C80	1.25×10^{1}	1×10^{-1}	8.0×10^{-3}	
		The Chip		
Vehusheia et al.	0.046	1.5	32.6	[122]

In contrast to the industrial microcalorimeters, the microfluidic chip presented in this chapter prioritizes system flexibility by eliminating additional bulky components. This is the main trade-off in regards to the detection limit of the system. Previous chip calorimeters also rely on an array of different additional thermally stabilizing components. As shown, the measurements are performed in laboratory environments, suggesting tabletop applications of the calorimetric microfluidic chip. In addition, the PDMS chip fabrication aspect allows for versatility in the channel design possibilities. Further, in the chip the thermal power of an *E. coli* bacteria is measured in agreement with literature values. The straight-forward differential compensation scheme allows for the resolution of the exponential growth curve of the bacteria, by common-mode rejection of the background thermal fluctuations. A versatile calorimetric microfluidic chip combining the benefits of the versatility of microfluidic chip fabrication with calorimetric sample measurements is shown. This system could be applied to different lab-on-chip systems [103, 104, 134] for direct information about metabolic changes. Further, it could give insight in microbiology on the investigations of cell response to environmental change, drug delivery efficiency, gene expression, and exothermic reactions [135, 136].

To relate the measured optical density to a biofilm, the thermal power densities are approximated and compared in Figure 5.16. A biofilm contains 15% active bacteria [137]. Assuming 3.5 pW as the bacterial thermal power of *E. coli* and an individual bacterial volume of 1.49 μm^3 [138], a thermal power density of 350000 W/m^3 is determined for a bacterial biofilm. Comparing this to the limit of detection determined of the microfluidic chip of 1707 W/m^3, the biofilm is orders of magnitude higher. Thereby, the microfluidic chip in its current configuration shows the possibility of measuring the onset of a growing bacterial biofilm using heat flux measurements.

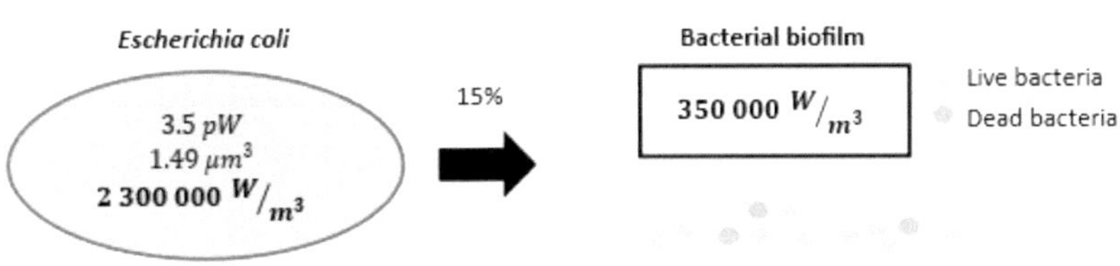

Figure 5.16: Calculation of the thermal power density of a biofilm. Open Access figure adapted from [122] published under open access under CC BY license (Creative Commons Attribution 4.0 International license).

5.2 Double Channel Microfluidic Chip

A variation of the one channel level microfluidic chip is the double channel microfluidic chip as illustrated in Figure 5.17. The fabrication process is as described in chapter 4.1.2 with channel height of 320 µm, width of 12 mm, and length of 12 mm. The purpose of the second top channel is to emulate the thermal environment of the blood flow in the aorta. This is achieved by calculating the heat transfer coefficient in the aorta and matching it in the microfluidic channel. The bottom channel is there to emulate a growing infectious environment of the vascular graft using an exothermic chemical reaction. This chapter is based on work that has previously been published [123].

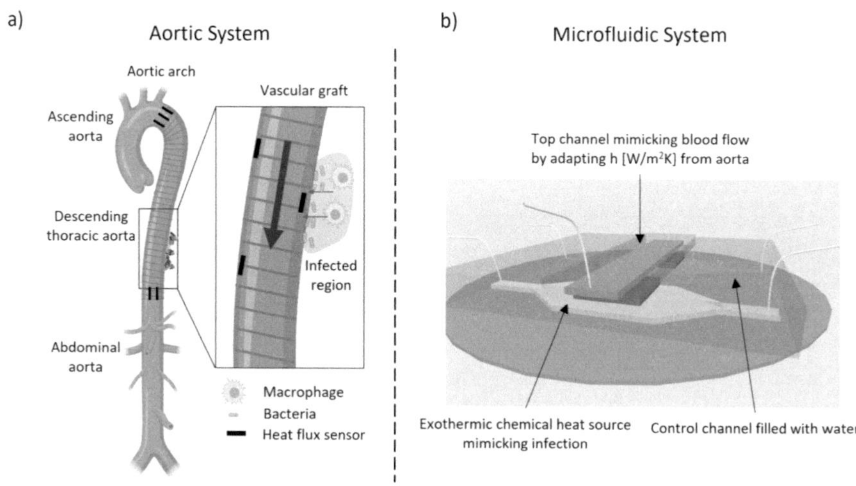

Figure 5.17: Schematic overview over the double channel microfluidic chip towards the emulation of the thermal environment of the aortic system. The microfluidic chip is based on the same design as the one channel level microfluidic chip. Figure adapted from [123] published under the Creative Commons Attribution 4.0 International license. The aortic vascular graft sketch is drawn using BioRender.com.

As presented in the one level microfluidic chip, there are two channels over which two heat flux sensors are placed. One channel is the reference channel filled with DI water and the other channel is filled with the sodium hydroxide (20 mL of 0.5 M) catalyzed hydrolysis of methyl paraben (152.32 mg) [127], respectively at 98.8 µL/min. In addition to these two channels, the top channel is also filled with DI water as shown in Figure 5.17.

In mimicking the thermal environment of the vascular graft, the heat transfer coefficient of the top channel matches the calculated heat transfer coefficient of the vascular graft. Further, the material of the channel is PDMS with a thermal conductivity of k_{PDMS}=0.15 W/mK which is lower than the thermal conductivity of standard aortic vascular graft materials such as ePTFE, k_{ePTFE}=0.35 W/mK, PTFE, k_{PTFE}=0.20 W/mK, and Dacron (PETP), k_{PETP}=0.51 W/mK. This means that the mircrofluidic chip system is a lower limit system and the heat transfer efficiency is higher for standard aortic vascular graft materials (also discussed in chapter 6.2.1). The bottom channel filled with the exothermic chemical reaction emulates the heat flow of a bacterial infection. The heat flow of the exothermic chemical reaction decreases exponentially which is exactly the opposite to the expected behavior of bacterial growth which increases exponentially and the thermal power output covered is in the similar range as the thermal output of bacterial growth. In literature, pulsatility has previously been shown to have negligible influence on the heat transfer coefficient [114], however, this is further discussed in detail in chapter 6.1.1.2.

To emulate the thermal conditions of the aortic flow in the upper channel of the microfluidic chip, a comprehensive analysis of both the aortic flow and upper channel flow environments was conducted, as shown in Figure 5.18.

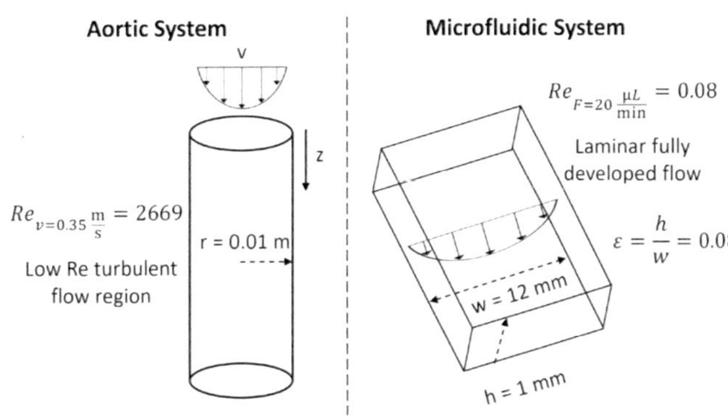

Figure 5.18: The aortic system is in the low Reynolds number turbulent flow region with $Re_{aorticsystem}$=2669. The microfluidic system of the top channel is in the laminar fully developed flow with $Re_{microfluidicsystem}$=0.08. Figure adapted from [123] published under the Creative Commons Attribution 4.0 International license.

The heat transfer coefficient of the aortic system is determined theoretically based on the set of equations presented by Mersmann [119] which is described in chapter 3.3.2.4 for the low Reynolds number turbulent flow regime in a cylinder. With this equation, the heat transfer coefficient is determined in dependence of the distance travelled from the inlet, z (as discussed in chapter 3.3.3). Figure 5.19 a) shows that for the case of the aortic vascular graft system, the heat transfer coefficient decreases with increasing distance travelled from the inlet, z, and it plateaus at 800 W/(m^{2}K) at z = 0.20 m. The relevant heat transfer coefficient region for the vascular graft is shown with a blue background .

In comparison, the microfluidic chip top channel, as shown in Figure 5.19 b), also shows a decreasing heat transfer coefficient with an increasing ϵ (channel thickness to width ratio). As described in chapter 3.2.6 and 3.1.5, L_h = 3.1 µm and L_{th} = 14.5 µm as shown in the Appendix in Table 5.2 (F=20 µL/min = 3.333×10^{-10} m^3/s), and the flow is laminar with Re = 0.08. Thereby, the heat transfer coefficient is determined as discussed in chapter 3.3.3. With a fabricated microfluidic chip of 1 mm thickness and 12 mm width, the relevant heat transfer coefficient is 4750 W/(m^{2}K) and it is shown with a pink line in Figure 5.19 b). A heat transfer coefficient of 4750 W/(m^{2}K) corresponds to z = 1 cm along the entrance of the vascular graft inlet.

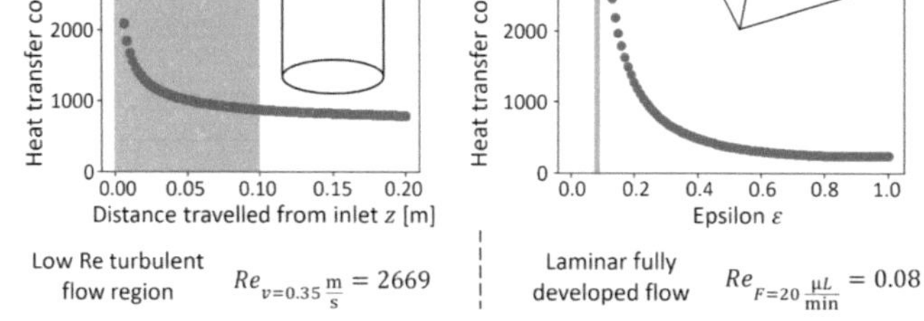

Figure 5.19: a) the aortic system is in the low Reynolds number turbulent flow region with $Re_{aorticsystem}$=2669. b) the microfluidic system of the top channel is in the laminar fully developed flow with $Re_{microfluidicsystem}$=0.08. Figure adapted from [123] published under the Creative Commons Attribution 4.0 International license.

Two different experiments were conducted in the double channel microfluidic chip as shown in Figure 5.20. One in which the top channel is filled with flowing water and another where the channel is filled with air. Prior to both experiments, a calibration experiment is conducted where the baseline is determined for the differential compensation scheme. The heat flux values plotted in Figure 5.20 is the differential heat flux determined by the analysis between the two channels as both described in chapter 5.1.2 and in literature [122].

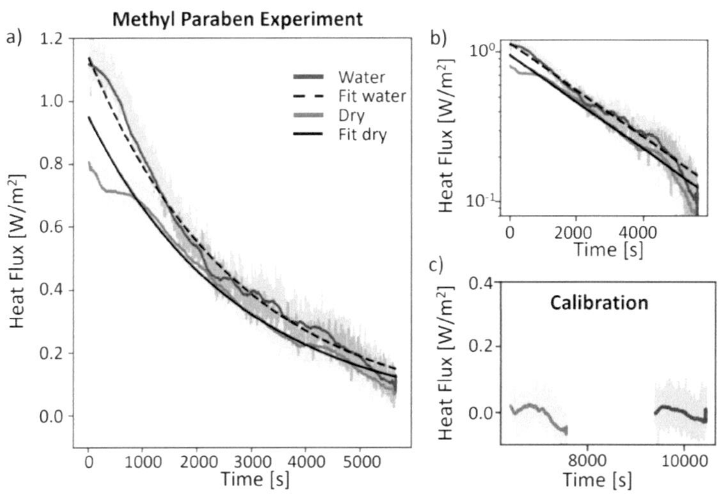

Figure 5.20: a) the differentially measured heat flux for water flowing on the top channel and also for air in the top channel upon addition of the exothermic chemical reaction. b) the exponential fit of the differentially measured heat flux with methyl paraben. c) both bottom channels are filled with DI water in the calibration phase with water or air in the top channel. Figure adapted from [123] published under the Creative Commons Attribution 4.0 International license.

There is a difference in the heat flux measured in the case where the top channel is filled with flowing water and where the top channel is filled with air. In both cases one of the bottom channels is filled with the exothermic chemical reaction. Both of the experiments show a heat flux decay rate of -3.6×10^{-4} 1/s, which fits the literature value -3.4×10^{-4} 1/s for experiments at 23°C [127]. This allows for the conclusion that the measured heat flux is originating from the base catalyzed hydrolysis of methyl paraben. The heat flux at t = 0 s are q_{dry}=0.95 W/m^2 and q_{flow}=1.14 W/m^2 for the respective channels. In comparing these two values, a ratio is determined as $B_{exp}=q_{flow}/q_{dry}$=1.20. A higher heat flux in the case of water flow through the top channel is to be expected as the convective heat transfer (water flow acts as a heat drain) presents a lower thermal resistance than conductive heat transfer to a non-moving air layer. This allows for the determination of the fact that an additional top channel filled with flowing water will increase the heat flux of a signal by 20%.

5.3 Thermally Fluctuating Environment

A variation in the experimental environment of the one channel level microfluidic chip experiments as described above, is the thermally not stabilized experimental environment (as described in chapter 4.2.2.2). In this chapter a microfluidic chip placed in a thermally fluctuating environment with a smaller heat flow source (channel size) and smaller sensor (decreased sensor sensitivity) is explored. The purpose of the thermally not stabilized environment is to further emulate the thermally fluctuating environment of an aortic vascular graft. The temperature surrounding an implant is affected by regular biological and external fluctuations (sleep cycles, fever, sports, and fluctuating weather). The sensor with the worse resolution is of smaller size, which would effectively allow for the integration in the aortic vascular graft polymeric wall. The measured heat flux in the bacterial results is compared to the results of the biological experiments presented in chapter 5.1.2. This section is based on previously published work [125, 126].

The chip fabrication process is as described in chapter 4.1.1, with channel height of 170 µm, width of 6 mm, and length of 6 mm, and a heat flux sensor gSKIN XP (resolution: 0.06 W/m^2, datasheet in Appendix in chapter C.1). The experimental environment of the thermally fluctuating experiment is a temperature controlled room without any further temperature stabilizing elements (PMMA box and copper piece) as described in chapter 4.2.2.2.

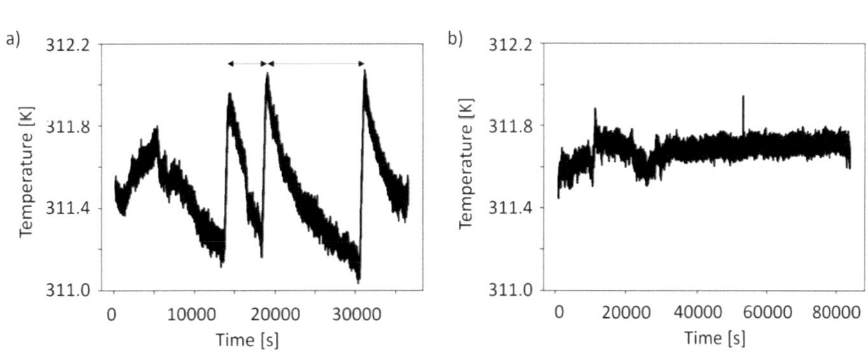

Figure 5.21: Temperature measurements throughout the whole experiment in both the thermally not stabilized (a) and the thermally stable environment (b). The fluctuations in a) are in the range of 1 K from 1.0 h to 3.3 h, whereas in b) the range of temperature is 0.3 K over 4.2 h. Open Access figure adapted from [125] published under open access under CC BY license (Creative Commons Attribution 4.0 International license).

The temperature measured in the thermally not stabilized and thermally stable environments is shown in Figure 5.21 a) and b), respectively. The temperature in the microfluidic chip placed in a 37°C room fluctuates 1 K over 3.3 h regularly as shown in Figure 5.21 a), influencing the heat flux signal at each temperature peak. In the thermally stabilized system (PMMA box and copper), the temperature fluctuates 0.3 K over 4.2 h as shown in Figure 5.21 b). Both of these temperature measurements were done with a PT1000 which is embedded in PDMS and placed in proximity to the microfluidic channel. The same differential heat flux compensation scheme as previously used is used for the two microfluidic channels (control and bacteria) as illustrated in Figure 5.22.

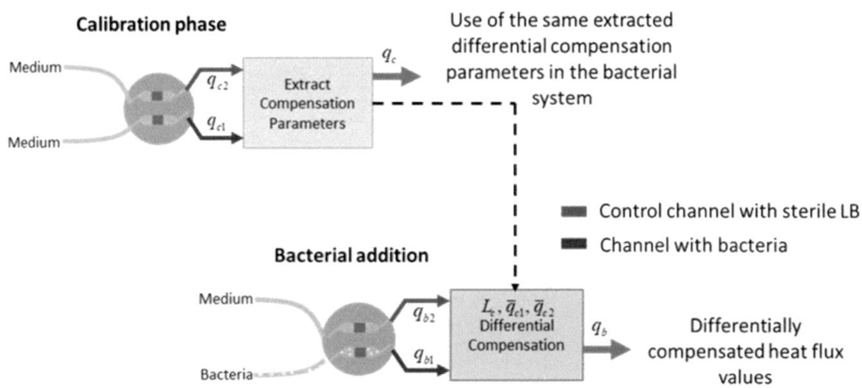

Figure 5.22: A schematic overview over the differential compensation scheme as previously shown [122]. In the calibration phase, the differential compensation parameters are extracted and subsequently used when the bacteria are added. This allows for the extraction of the heat flux signal by the bacterial growth without the thermal background fluctuations.

Figure 5.23 shows the calibration phase of the thermally not stabilized experiment where both channels are filled with sterile LB media. a) shows the corrected signal (according to the data analysis in chapter 5.1.2.3) from both channels and b) shows the differentially compensated heat flux signal of the calibration phase. The peak in the heat flux corresponds to peak in the temperature of the temperature-controlled room as shown in Figure 5.21.

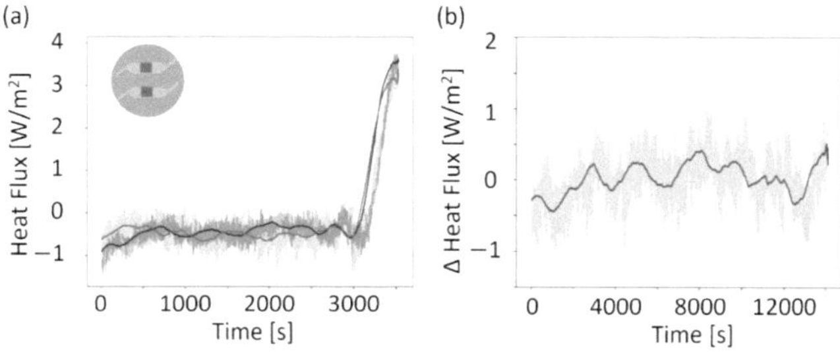

Figure 5.23: Raw data and the differentially compensated heat flux values in the calibration phase. a) shows the data during the calibration phase and b) shows the data after the differential compensation between the two systems. The raw data is shown in the opaque color and the data averaged over 200 data points is shown in the darker color. Open Access figure adapted from [125] published under open access under CC BY license (Creative Commons Attribution 4.0 International license).

Figure 5.24 shows the heat flux data upon addition of bacteria. a) shows the heat flux signal of both sensors where bacteria is added in the blue signal. The point at which the heat flux increases due to the bacterial growth is indicated with an arrow in both a) and b). Again the differentially compensated heat flux signal is shown. However, comparing Figure 5.23 b) and Figure 5.24 b) it is difficult to clearly identify a shift in the heat flux signal measured upon the addition of bacteria.

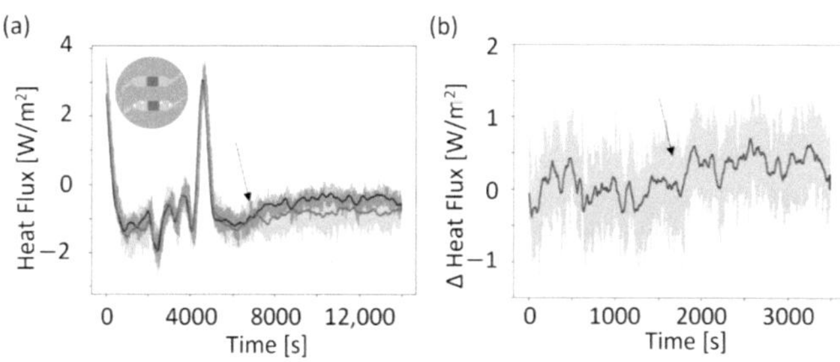

Figure 5.24: Raw data and the differentially compensated heat flux values. The heat flux data shown is upon the addition of bacteria in a) and b). An increase in the heat flux can be identified in d) at around 8000 seconds. The blue data is of the channel in which bacteria is added and the red data is of the calibration channel with only LB media. Open Access figure adapted from [125] published under open access under CC BY license (Creative Commons Attribution 4.0 International license).

To clearly visualize an shift of an increased heat flux signal upon addition of bacteria, a cumulative distribution function is made to compare the calibration phase with two different phases in the data in Figure 5.24 b). The two phases indicated by the arrow at around 8000 s and two horizontal lines in Figure 5.25 a). The horizontal lines are named region 1 (t = 0 to 7500 s) and region 2 (t = 7500 to 13000 s). Figure 5.25 b) shows the cumulative distribution function in the three different cases: calibration, region 1 and region 2. As is clearly shown, the calibration phase and region 1 show the same cumulative distribution function as the calibration phase whereas the region 2 shows a shift.

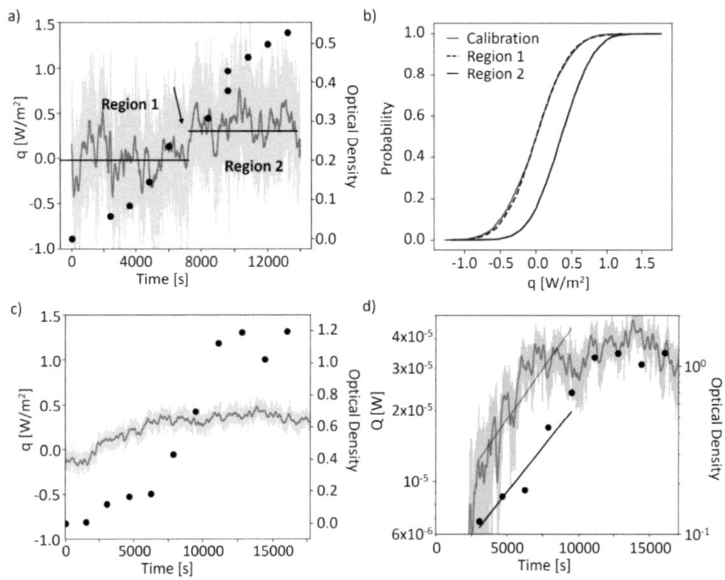

Figure 5.25: a) shows the optical measurement in parallel to the differential heat flux data shown in Figure 5.24. b) shows the cumulative distribution function comparing the calibration phase with the two heat flux phases shown in a). c) and d) shows the heat flux data shown in Figure 5.10. Open Access figure adapted from [125] published under open access under CC BY license (Creative Commons Attribution 4.0 International license).

Figure 5.25 a), c) and d) also shows the optical density measurements of both the thermally not stabilized environment as shown in a) and for the thermally stabilized environment as shown in c) and d). The heat flux values shown in c) and d) are the same as presented in chapter 5.2. The comparison shows how in the thermally stabilized case, the exponential growth of the bacteria is resolvable, whereas in the thermally not stabilized environment, only a step-like increase in the heat flux value is identifiable with increasing bacterial population. The step-wise increase of the heat flux is at optical density of 0.3.

A comparison between the thermally not stabilized and thermally stabilized system is shown in Table 5.4. The detection limit in terms of cells is determined by the optical density expressed in cells/mL multiplied by the channel volume. The detection of bacterial growth in the thermally not stabilized environment is confirmed through a comparison of the cumulative distribution function comparing the calibration phase to the point at which the bacterial growth is detected. However, it is only able to detect whether or not a bacterial colony has grown, and is not able to follow the exponential growth curve of the bacterial colony as in the thermally stabilized system.

Table 5.4: An overview over the different system parameters of the thermally not stabilized system and the thermally stabilized system.

Property	Thermally Not Stabilized	Thermally Stabilized
Growth detection	Yes	Yes
Channel size (underneath sensor)	6 mm × 6 mm × 170 µm	12 mm × 12 mm × 320 µm
Channel volume	6 µL	46 µL
Sensor	gSKIN XM	gSKIN XP
Sensor resolution	0.41 W/m^2	0.06 W/m^2
Temperature fluctuation	1 K	0.3 K
Standard deviation heat flux	0.32 W/m^2	0.05 W/m^2
Standard deviation averaged heat flux	0.20 W/m^2	0.02 W/m^2
Detection limit OD	3×10^8 cells/mL	2×10^7 cells/mL
Detection limit cells	1.8×10^6 cells	9.2×10^5 cells

In the thermally stabilized system, the limit of detection is established based on the determined sensitivity and noise as described in the Appendix in chapter E.5. However, this is not the method applied to the thermally fluctuating experiment since it is not possible with the given data. In this method, the limit of detection is determined in relation to the optical density (bacterial population) that produces a detectable increase in the heat flux. This is statistically confirmed using the cumulative distribution function as shown in Figure 5.25 b). This solely allows for the determination of a heat flux threshold at which a bacterial colony is detected. In comparison, the thermally stabilized system allows for the resolution of the exponential growth phase as shown in Figure 5.13. Due to the different modes of determining the limit of detection in the thermally not stabilized and thermally stabilized systems, the two limit of detections for thermal power density are not directly comparable.

With this system, it is shown that it is possible to detect a growing bacterial colony even in a thermally not stabilized environment and also with a heat flux sensor resolution of 0.41 W/m^2 in comparison to 0.06 W/m^2. Comparing the limit of detection in terms of bacterial population, the amount detected corresponds to implant bacterial seeding amounts of 2.0×10^6 and 6.2×10^6 bacteria for implant infection research [139–141]. This and in addition to the thermally fluctuating environment shows the feasibility of early detection of infection on aortic vascular implants by experiments with an *in vitro* microfluidic chip.

5.4 Immune Cell Experiments

As previously discussed, bacteria are not the only source of heat flow at an infected implant. Immune cells and especially their immune response towards bacteria are another source of heat flow. For that purpose, THP-1 cells as monocytes as well as differentiated macrophages were analysed by flow cytometry/fluorescence activated cell sorting (FACS). After determining the cell type through FACS, the thermal power originating from the cells is measured using microcalorimetric measurements. Following these experiments, the integration of immune cells in a microfluidic chip is explored.

5.4.1 FACS Experiments

The THP-1 cells are stimulated according to chapter 4.5.1 and prepared for FACS as described in chapter 4.5.3. The plate prepared for FACS analysis is shown in Figure 5.26. Different THP-1 stimulation pathways (as shown in Figure 4.16) are compared: THP-1 monocytes, macrophages stimulated with PMA at different time points, macrophages stimulated with PMA and LPS/IFN-γ, different concentrations of *E. coli* (gram negative) bacteria, different concentrations of *E. faecalis* (gram positive) bacteria, and stimulation pathways within the microfluidic chip channel. PMA stimulated THP-1 cells are used for fluorescent compensation in addition to beads. The cells are detached from the wells using Accutase as discussed in chapter 4.5.2.

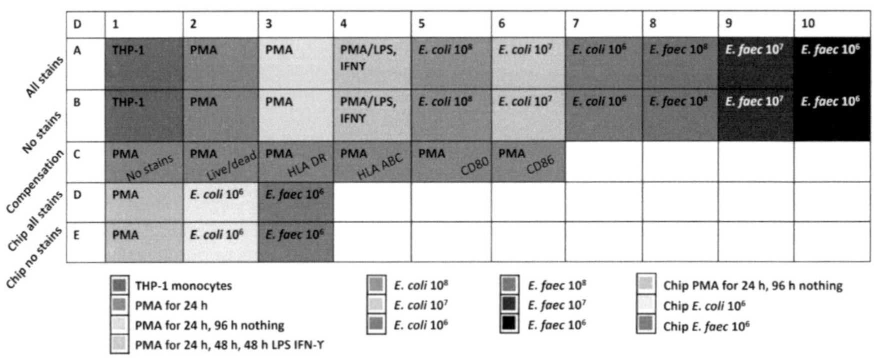

Figure 5.26: FACS plate overview. Color coding in accordance to the following figures below.

The data shown in Figure 5.27, 5.28, and 5.29 is of the live cells. Dead cells are not considered in the data analysis.

Monocytes and macrophages stimulated with PMA and LPS/IFN-γ are compared in Figure 5.27. The antigens and the costimulatory molecules HLA ABC, HLA DR, CD80, and CD86 are compared. HLA is the human leukocyte antigen which is a human major histocompatibility complex (MHC). HLA-ABC antigens are used to show MHC of class I expression in the immune cells. No significant up- or downregulation of HLA-ABC is shown upon stimulation of THP-1 monocytes with PMA and PMA with LPS/IFN-γ as shown in Figure 5.27. HLA-DR antigens are molecules for MHC class II and are related to cell signalling and are upregulated in the case of increased signalling of the cell caused by increased stimulus. This is specifically upregulated upon an infection as is the case for M1 macrophages as illustrated in Figure 4.16. In Figure 5.27 b), PMA in addition to LPS/IFN-γ show significant upregulation of HLA-DR. However, the cells stimulated with PMA do not show any upregulation in relation to THP-1. CD80 and CD86 is upregulated for pro-inflammatory macrophages. In Figure 5.27 c) and d), only LPS/IFN-γ shows upregulation, suggesting only LPS/IFN-γ stimulated THP-1 cells are differentiated from monocytes to pro-inflammatory macrophages. In general, an increase in size and granularity of the immune cells mark the differentiation from monocytes to macrophages, only the antigen and costimulatory molecules give insight in whether or not the macrophages are also pro-inflammatory.

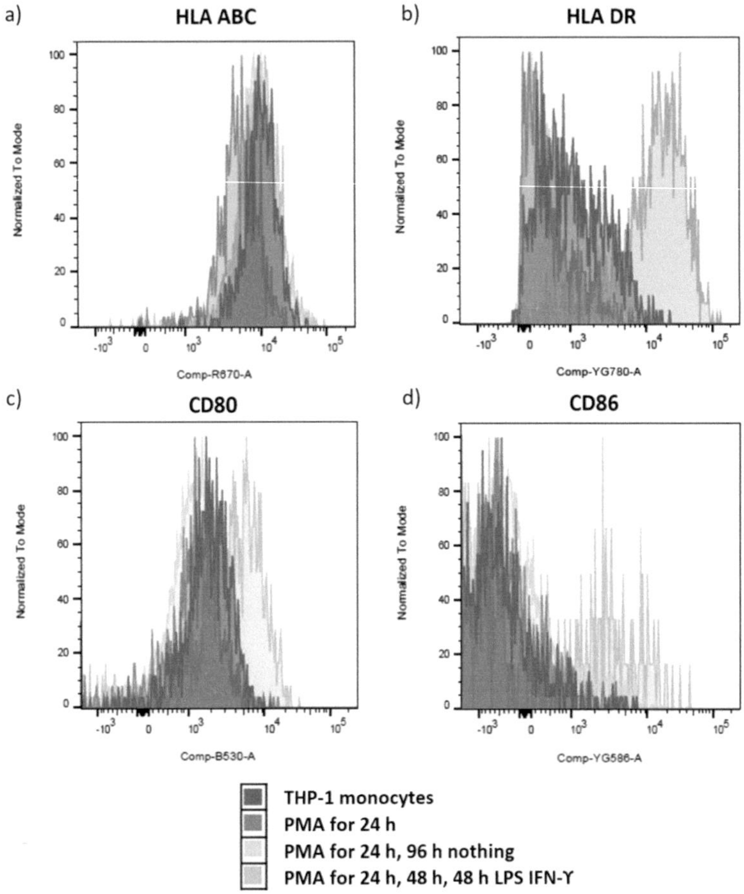

Figure 5.27: Pro-inflammatory macrophages derived from ThP-1 cells increase expression of HLA-DR, CD80, and CD86. The THP-1 cells were stimulated with PMA and LPS/IFN-γ. After measurement, the derived data was analysed using FlowJo. Shown are representative histograms of all groups.

A low amount of living cells were observed when THP-1 cells were treated with the two highest concentrations of *E. coli*, 10^8 and 10^7 cells/mL. When cells are treated with 10^7 *E. coli* cells/mL no difference is shown for HLA-ABC, HLA-DR and CD80 in comparison to unstimulated THP-1 cells (as shown when comparing Figure 5.27 and Figure 5.28). CD80 shows some downregulation for an *E. coli* concentration of 10^6 cells/mL. However, CD86 shows some upregulation similar to that of LPS/IFN-γ stimulated THP-1 cells.

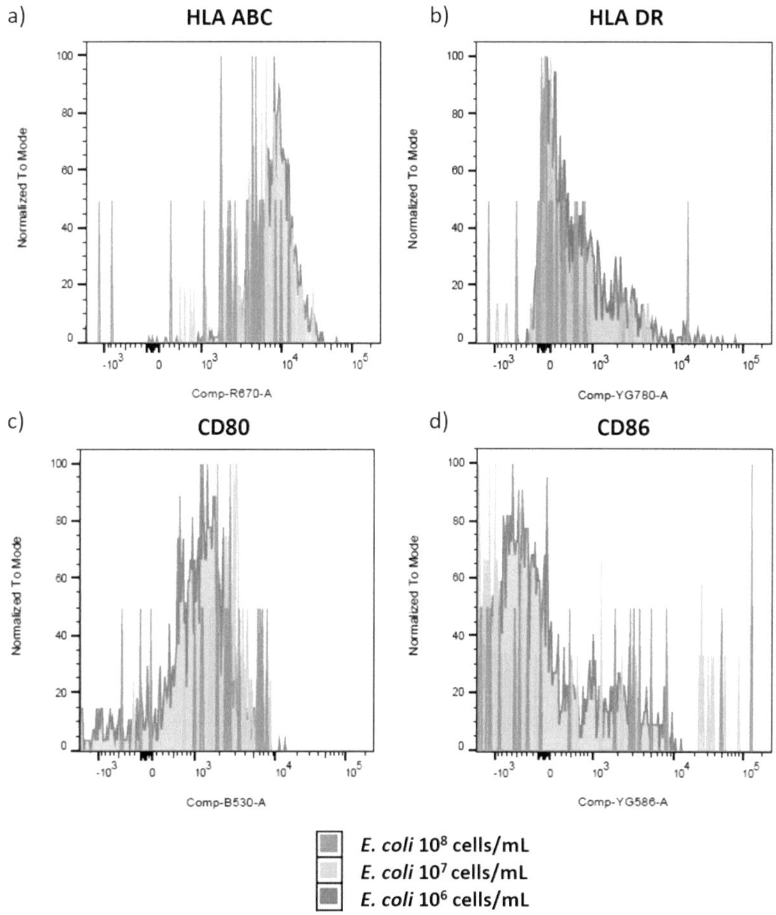

Figure 5.28: FACS data for 10^8, 10^7 and 10^6 cells/mL *E. faecalis* stimulated THP-1 monocytes.

Similarly to the stimulation by *E. coli*, stimulation by *E. faecalis* shows upregulation of CD86. Interestingly, the upregulation of *E. faecalis* increases with increasing bacterial cell concentration. This suggest concentration dependent THP-1 cell activation.

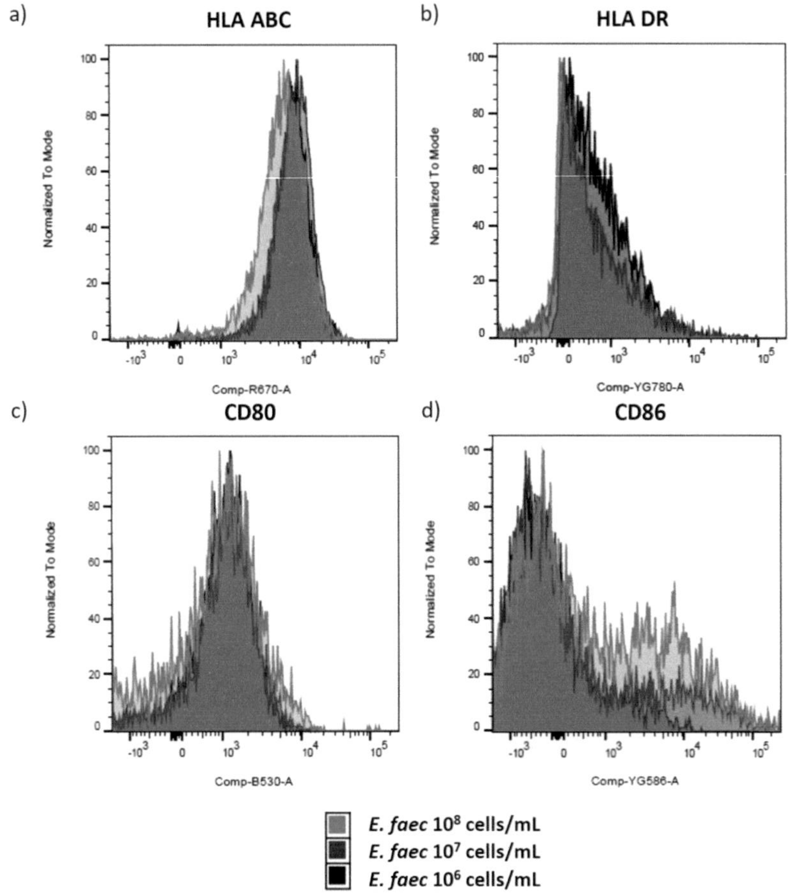

Figure 5.29: FACS data for 10^8, 10^7 and 10^6 cells/mL *E. faecalis* stimulated THP-1 monocytes.

Figure 5.30 shows the FACS data for samples stimulated in and detached from microfluidic chip. The cells removed from the chip show less upregulation of all of the antigens in comparison to stimulation in the wells. A slight upregulation of CD86 is still recognizable, however, it is significantly smaller than shown in Figure 5.27, 5.28, and 5.29. The most probable reason for less stimulation in the microfluidic chip is the lower amount of stimulants present in the channel. There is no precise control on whether or not all of the media is in fact replaced within the channel in addition to the occasional introduction of air bubbles.

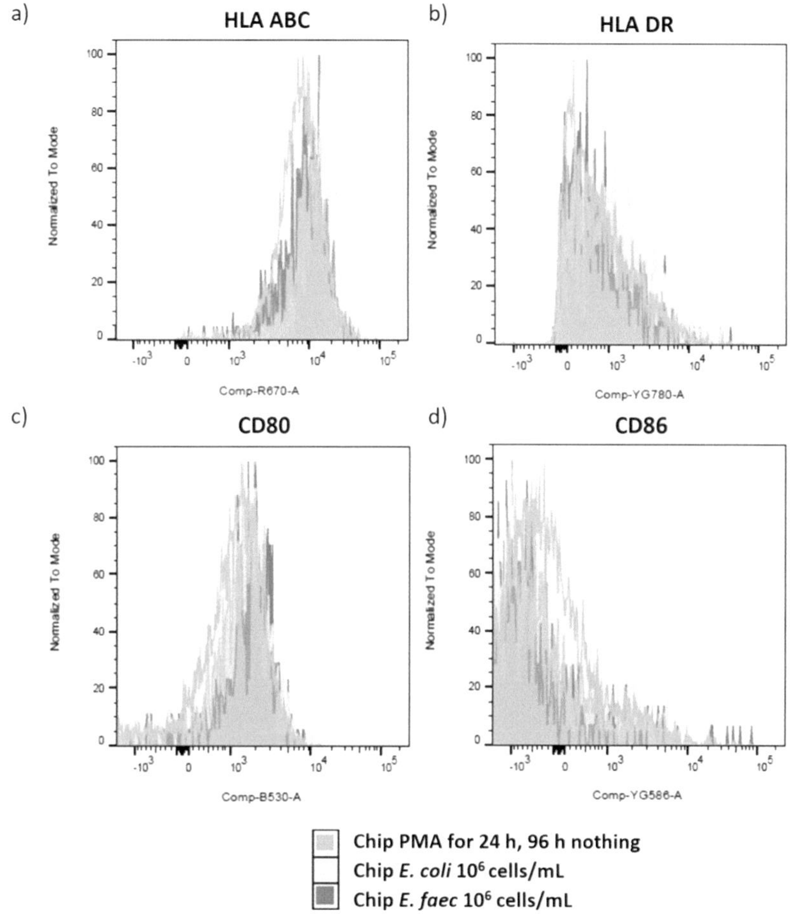

Figure 5.30: FACS data for the different stimulation in a microfluidic chip.

The percentage of living cells of each sample is shown in Table 5.5. In general, a high amount of live cells is found for cells removed by Accutase and stimulated in the wells. However, 10^8 cells/mL and 10^7 cells/mL *E. coli* only have 0.8% and 3.4% of live cells. The reason might be overstimulation or even a non-sterile *E. coli* vaccine. For microcalorimetric data, the percentage of live cells measured in FACS is used to calculate the thermal power per cell for samples removed by Accutase. As to be expected, due to the less ideal growing conditions and detachment procedure within the microfluidic chip, the percentage of live cells is consistently less than of the samples in the wells.

Table 5.5: An overview over the percentage live cells in each of the FACS samples.

Sample	Live cells
THP-1 Monocytes	96.1%
PMA for 24 h	93.0%
PMA for 24 h, 96 h nothing	85.1%
PMA for 24 h, 48 h, 48 h LPS IFN-γ	56.1%
E. coli 10^8 cells/mL	0.8%
E. coli 10^7 cells/mL	3.4%
E. coli 10^6 cells/mL	77.5%
E. faecalis 10^8 cells/mL	85.9%
E. faecalis 10^7 cells/mL	89.8%
E. faecalis 10^6 cells/mL	82.7%
Chip PMA for 24 h, 96 h nothing	67.3%
Chip *E. coli* 10^6 cells/mL	66.0%
Chip *E. faecalis* 10^6 cells/mL	50.4%

5.4.2 Immune Cell Microcalorimetric Experiments

5.4.2.1 Scraped Immune Cell Microcalorimetric Experiment

The microcalorimetric plate containing all the samples measured is shown in Figure 5.31. Two to four repetitions of each sample is placed in the plate. Each sample is stimulated as discussed in chapter 4.5.1. The two rows of water are for the thermal reference of the microcalorimetric system. The cells are collected by mechanical scraping from the wells instead of Accutase. 250 µL of 10^6 cells/mL sample is placed in each well, however, the total number of live cells is unknown in each sample. The samples are placed inside the microcalorimeter directly upon adding the stimulants to the cells. The heat flow measured is shown in Figure 5.32 and 5.33.

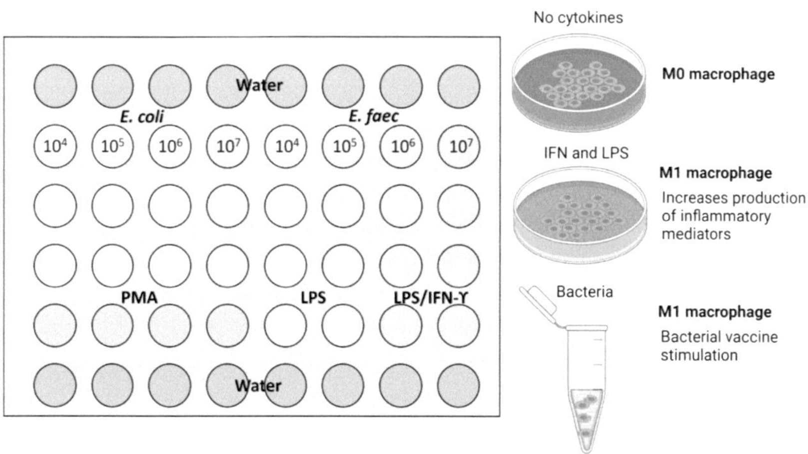

Figure 5.31: Microcalorimetric plate overview for the experiments shown in Figure 5.32 and 5.33. Sketch created with BioRender.com.

The heat flow measured for the PMA, LPS and IFN-γ is low as shown in Figure 5.32. The highest heat flow signal is measured for the PMA and LPS stimulated THP-1 monocytes at around 0.2 µW, and the lowest heat flow signal is for the cells only stimulated with PMA at around 0 µW. Assuming 250000 cells in each well, the heat flow per PMA and LPS stimulated cell is at 0.8 pW. However, high cell death is expected for cells removed by mechanical scraping.

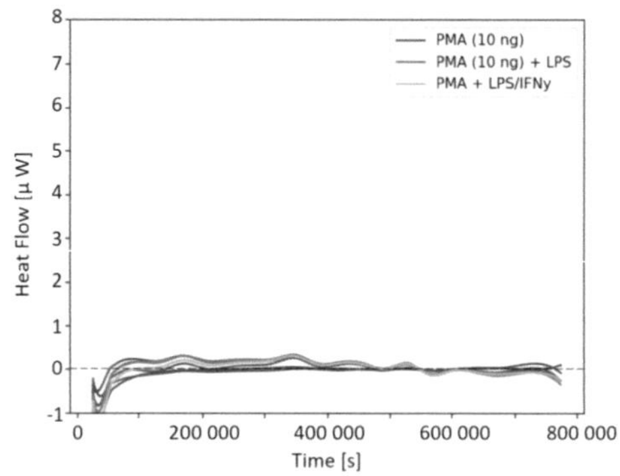

Figure 5.32: Microcalorimetric measurement of THP-1 cells stimulated by PMA, PMA and LPS, and PMA and LPS/IFN-γ. Cells removed by mechanical scraping.

Figure 5.33 shows the heat flow with stimulation by 10^4, 10^5, 10^6, and 10^7 bacteria/mL of *E. coli* and *E. faecalis*. The heat flow measured with stimuli of the bacteria is higher than stimuli by LPS and IFN-γ. For *E. coli* there is bacterial concentration dependence with increasing bacterial concentration decreasing the heat flow output. This suggest overstimulation of the immune cells. The *E. faecalis* show less indication of any trend regarding the heat flow output in relation to the varying concentrations.

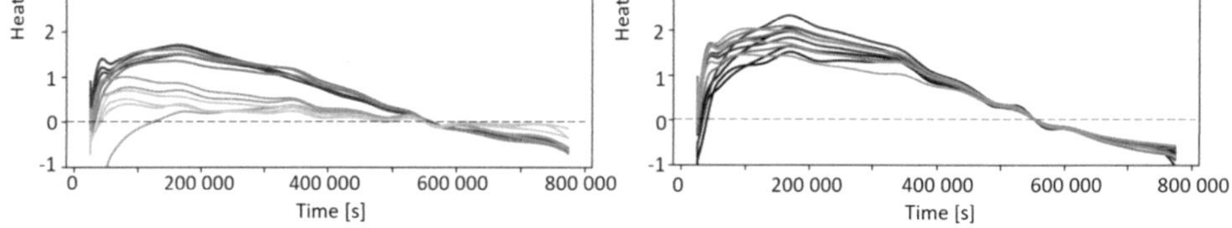

Figure 5.33: Microcalorimetric measurement of 10^4, 10^5, 10^6, and 10^7 bacteria/mL of *E. coli* and *E. faecalis* stimulating THP-1 cells. Cells removed by mechanical scraping.

5.4.2.2 Accutase Immune Cell Microcalorimetric Experiment

The microcalorimetric experiment is repeated with different bacterial concentrations and by using Accutase for cell removal. Thereby, the live cells in the samples can be estimated by the percentage of live cells measured in FACS as shown in Table 5.5.

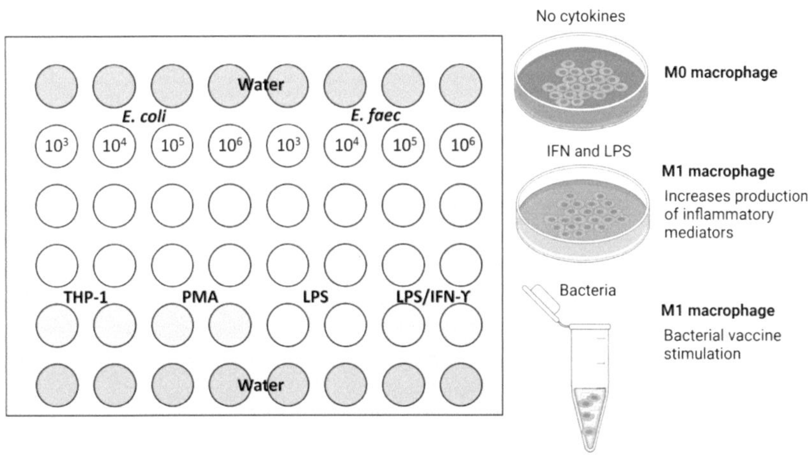

Figure 5.34: Microcalorimetric plate overview for the experiments shown in Figure 5.35 and 5.36. Sketch created with BioRender.com.

Similar to the heat flow measurements in Figure 5.32, the heat flow measured is less than the heat flow measured for immune cells stimulated by bacteria. However, the measured heat flow is larger than that of the cells collected by scraping, which is due to the larger amount of live cells.

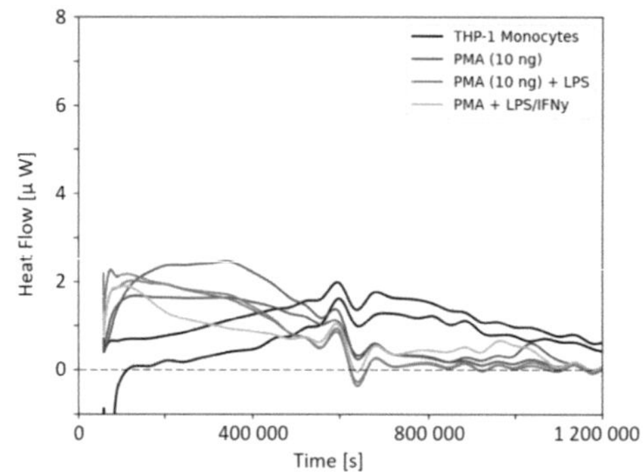

Figure 5.35: Microcalorimetric measurement of THP-1 cells and THP-1 stimulated by PMA, PMA and LPS, and PMA and LPS/IFN-γ. Cells removed by Accutase.

Similar to Figure 5.33, *E. coli* concentrations of 10^4, 10^5, and 10^6 decrease the measured heat flow with increasing bacterial concentration. However, the heat flux in the case of *E. faecalis* does not follow any specific order and the amount of living cells is larger.

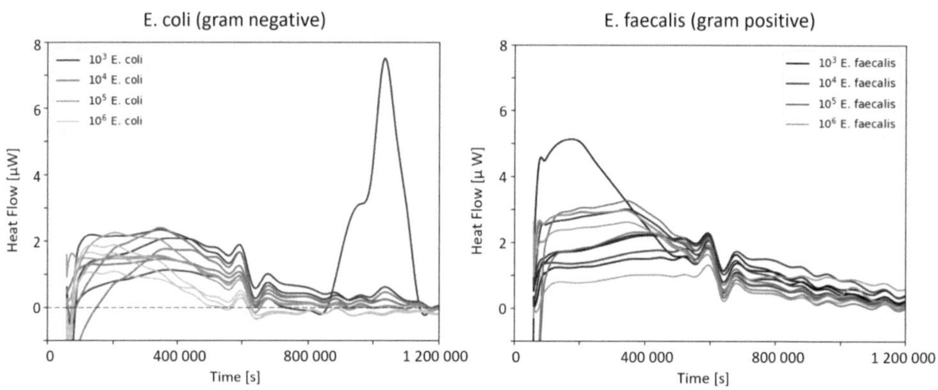

Figure 5.36: Microcalorimetric measurement of 10^3, 10^4, 10^5, and 10^6 bacteria/mL of *E. coli* and *E. faecalis* stimulating THP-1 cells (different concentrations to Figure 5.33). Cells removed by Accutase.

5.4.3 Immune Cells as a Heat Source

As shown in Table 5.5, the percentage of live cells of 10^6 *E. coli*/mL is 77.5%. The average thermal power achieved for a stimulation of 250000 immune cells with 10^6 *E. coli*/mL is 1.5 µW as shown in Figure 5.36. The bacterial stimulation of the immune cell is achieved with dead *E. coli* cells through the vaccine protocol described in Table 4.7, therefore only the immune cells contribute to the measured heat flow by the microcalorimeter (dead bacteria do not create heat flow). With an amount of live immune cells of 193750 (77.5% of 250000 cells), the heat flow per immune cell is 7.7 pW which is higher than that for *E. coli* cells of 3.5 pW [130].

For immune cell stimulation by 10^6 dead *E. faecalis*/mL, the average heat flow measured is 3 µW. The measured live cells is 82.7% as shown in Table 5.5. Therefore the total live and stimulated heat flow originating from immune cells is 206750. The heat flow per immune cell is thereby 14.5 pW, which is also higher than that for *E. coli* cells of 3.5 pW [130].

In literature there is limited information regarding the microcalorimetric measurements of THP-1 cells, especially regarding bacterial stimulation. In the microfluidic chips discussed (chapter 5.1.2.3 and 5.3), the heat flow by bacterial growth is measured and a limit of detection regarding the bacterial cell population is determined. However, at an infected implant, the immune cells also contribute to the localized heat flow increase, and each individual immune cell has a higher heat flow per cell than bacterial cells. Therefore, even lower bacterial concentrations can be detected than the determined LODs in Table 5.4 in the microfluidic chips. This further supports the feasibility of early detection of infection on aortic vascular graft implants using heat flux sensing.

6 FEM Simulations

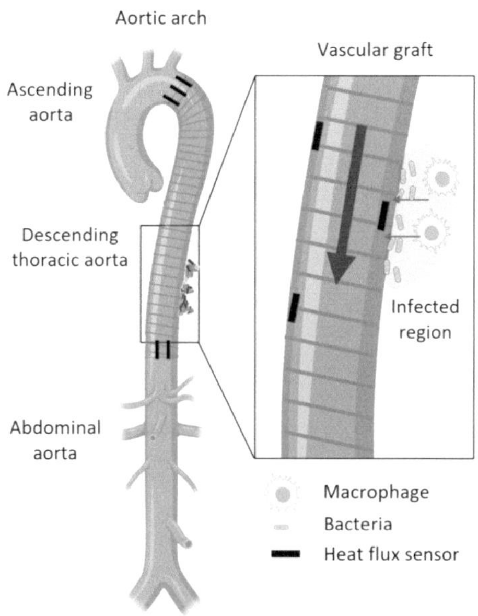

Figure 6.1: Overview over the finite element modelled system. Open Access figure adapted from [1] published under open access under CC BY license (Creative Commons Attribution 4.0 International license). Sketch created with BioRender.com.

In the previous chapters, the experimental results of different *in vitro* calorimetric microfluidic chip systems mimicking infectious growth on implant surfaces have been discussed. It is shown that it is possible to detect the onset of the exponential growth curve of bacteria with the *in vitro* chip (also in thermally not stabilized environments) using heat flux sensing. Different heat flux values measured correspond to a certain amount of bacteria. The onset of exponential growth is postulated to correspond to the early growth phase of bacteria on implant surfaces, as there currently is no clinical literature on the bacterial growth curve on *in vivo* implant infections. With the experimental values in mind, further investigations of the use of heat flux sensors for the detection of infectious growth in the *in vivo* system of the aortic vascular graft using finite element modelling are done. In the following chapters, the aortic vascular graft system is investigated as previously published in literature [1]. This system theoretically investigates the complete implementation of a heat flux sensor for the early detection of infection on a vascular graft as illustrated in Figure 6.1. The investigation starts with a fluid dynamic modelling of pulsatile flow and determination of the heat transfer coefficient at the cylinder wall. Parametric sweeps with heat transfer simulations is performed to investigate the measurable heat flux of growing infections for the determination of the limit of detection. A color version of this chapter is available at https://doi.org/10.3929/ethz-b-000703739.

6.1 Fluid and Thermodynamic Investigations of Aortic Vascular Graft

As shown in Figure 6.1, the descending thoracic aorta is the region of interest. An infection is assumed on the tissue side, and blood flows through the aortic vascular graft implant. The fluid dynamic aspects of the blood flow are modelled using COMSOL and the heat transfer coefficient at the implant wall is calculated. The heat transfer coefficient is dependent on the blood pulsatility, graft radius, distance travelled from inlet, and flow region. In literature, not many experimental studies are to be found on pulsatile flow and its influence on the heat transfer coefficient. However, in literature different numerical studies show different influences (frequency, distance travelled from inlet, and hydrodynamic and thermal development phase) on the heat transfer coefficient through pulsatile flow, as discussed in the state of the art.

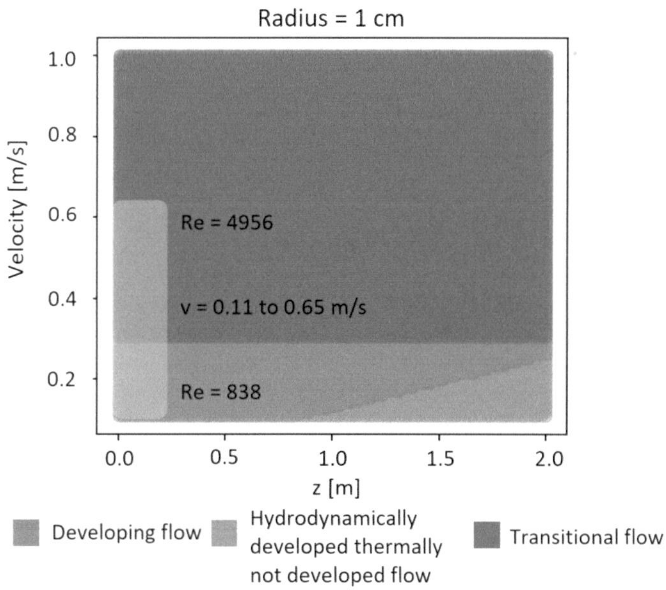

Figure 6.2: Flow regions of the aorta in the physiologically relevant region. Open Access figure adapted from [1] published under open access under CC BY license (Creative Commons Attribution 4.0 International license).

The flow environment within the aorta is defined according to chapter 3.3.2 in the theory chapter. The descending thoracic aorta is approached as a cylinder and assumes in general no Windkessel effect, Newtonian fluid behavior, and no aortic branching. The blood velocity is reported to range between 0.11 m/s to 0.65 m/s in the aorta [142] which gives an estimate average velocity, v_{avg}, of 0.35 m/s. As shown in Figure 6.2, the corresponding flow region is the developing flow region with $z < L_h$, L_{th} and the transitional turbulent low Reynolds number flow regime with $2300<Re<8000$. In general, the radius of the descending aorta is 1 cm [2], and the length is around 20 cm [143]. The corresponding flow environments relevant for aortic blood flow are shaded in white in Figure 6.2.

In the following chapters, three different types of flow are investigated: constant flow, sinusoidally oscillating flow, and flow from an interpolated pressure profile. The influence of the flow is investigated on the heat transfer coefficient. Subsequently, the range of heat transfer coefficients are applied in a heat transfer simulation allowing for the systematic investigation of the viability of infectious growth detection on a vascular graft implant.

6.1.1 Fluid Dynamic Simulation Results

As mentioned, the three different flow environments are constant flow, sinusoidally oscillating flow, and flow based on an interpolated pressure profile reported in literature [124]. The pulsatile flow simulations are based on the constant flow simulations through $v_z = v_{avg} + u$, where v_z is the total velocity in z direction, v_{avg} is the average aortic blood flow velocity, and u as the pulsatile element oscillating around 0 m/s. The the pulsatile blood flow is modeled as an additional component, however, the driving force of the flow is the average blood velocity v_{avg}. The COMSOL simulation setup is built using the Multiphysics module combining the low Reynolds $k - \epsilon$ turbulent flow (spf) and the heat transfer in solid and fluids module (ht), as described in detail in chapter 3.5.1.

6.1.1.1 Constant Velocity $\mathbf{u} = 0, \mathbf{v}_z = \mathbf{v}_{avg}$

The simulated velocity profile at increasing distances from the cylinder inlet, z, in the laminar flow regime show the development of the Hagen-Poiseuille velocity profile (Figure 6.3 a)). The given v_{avg} for laminar flow is 0.11 m/s with a Reynolds number of 838. Increasing v_{avg} to 0.35 m/s increases the Reynolds number to 2669, which is in the low Reynolds number turbulent flow regime. As the flow profile develops along the distance travelled from the inlet, z, the velocity profile shows a flat plateau instead of a parabolic Hagen-Poiseuille velocity profile indicating the turbulence present in the flow.

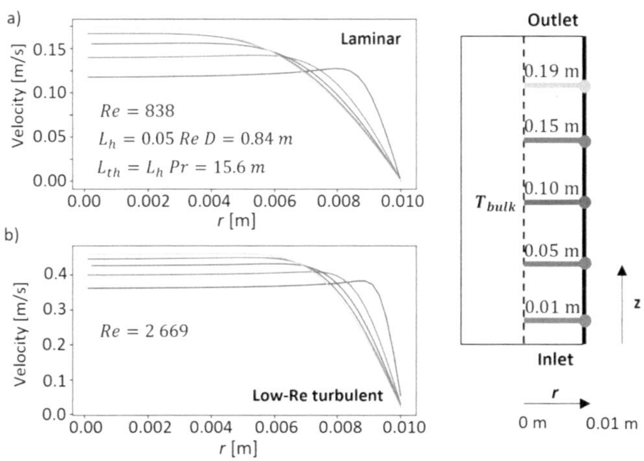

Figure 6.3: Velocity profiles at different points along the direction of flow in the laminar and low Reynolds number turbulent flow regions. Open Access figure adapted from [1] published under open access under CC BY license (Creative Commons Attribution 4.0 International license).

The heat transfer coefficient, h, is calculated by the given heat flux, q, at the implant wall using the local temperature at the implant wall, T_{local}, and the temperature in the cylinder, T_{bulk}, by $h = \frac{q}{T_{local} - T_{bulk}}$ (described in chapter 3.5.2). Similar to what is reported in literature, the heat transfer coefficient in Figure 6.4 decreases with increasing distance from the fluid inlet in a cylinder [109]. Furthermore, the heat transfer coefficient increases with increasing velocity.

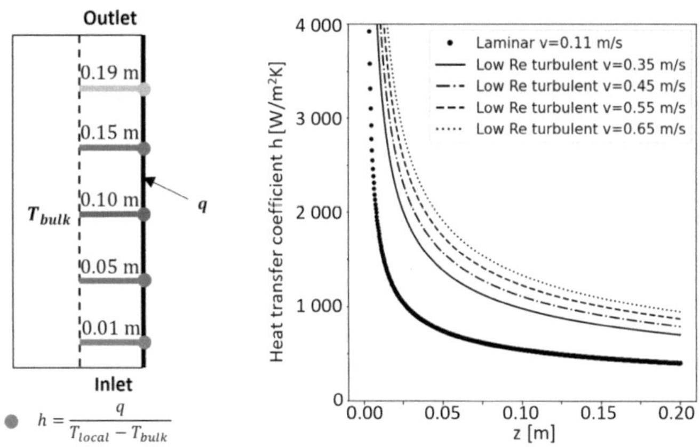

Figure 6.4: Velocity profiles at different points along the direction of flow in the laminar and low Reynolds number turbulent flow regions. Open Access figure adapted from [1] published under open access under CC BY license (Creative Commons Attribution 4.0 International license).

There is a steep decrease in the heat transfer coefficient from $z = 0.00$ m to 0.05 m and from 0.05 m there is a gradual slope. The total slope of the heat transfer coefficient in the low Reynolds number turbulent regime can thereby be captured by a power law function:

$$h(z) = az^{-0.5} + cz^{-1.5} \tag{6.1}$$

with a and c as fitting parameters. Figure 6.5 shows the comparison between the exponential function and the simulated heat transfer coefficients. The R^2 values of the fits for the different velocities range between 0.96 to 1.00. With this function, a predictability of the heat transfer coefficient in the low Reynolds number flow region in relation to the distance travelled from the inlet, z is found. The fitting parameters a and c show a linear relationship with the fluid velocity as discussed in the Appendix in Chapter F.1.1.

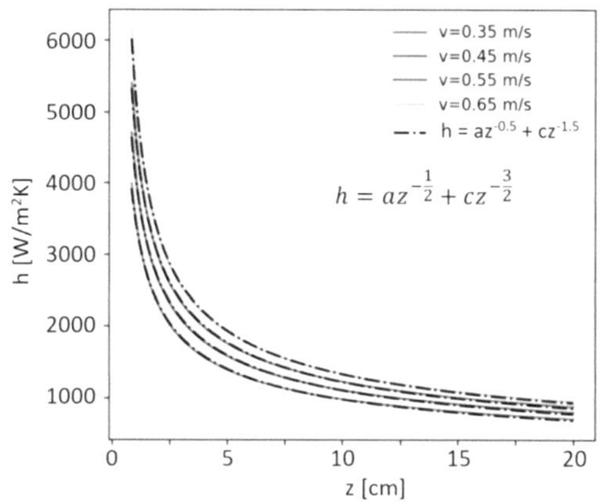

Figure 6.5: Power law fit of the heat transfer coefficient along the aortic wall in the z direction. Open Access figure adapted from [1] published under open access under CC BY license (Creative Commons Attribution 4.0 International license).

Although vascular grafts are not tapered, the aortic radius decreases along the direction of blood flow [144]. Figure 6.6 shows that the heat transfer coefficient increases with decreasing radius.

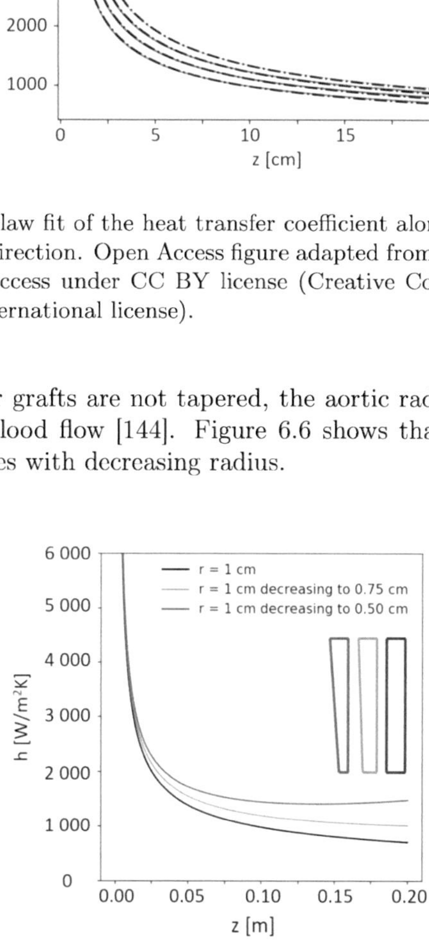

Figure 6.6: Influence of radius variations on the heat transfer coefficient. Open Access figure adapted from [1] published under open access under CC BY license (Creative Commons Attribution 4.0 International license).

6.1.1.2 Pulsatile Velocity $\mathbf{v}_z = \mathbf{v}_{avg} + \mathbf{u}$

The pulsatile element of the blood flow is introduced by u through a partial differential equation as described in the theory chapter 3.5.1. The partial differential equation solves for a velocity with a given sinusoidal pressure input with this equation $\frac{1}{c}\frac{\partial P}{\partial t} = -Asin(2\pi f(t - \frac{z}{c}))\frac{1}{c}$, where c is the pulse wave velocity in blood, p is the pressure, A is the sine wave amplitude, f is the sine wave frequency, and t is the time. The full flow profile is shown in the Appendix in Figure F.2. The simulated heat transfer coefficient with an input sinusoidal pulsatile velocity element with f = 1 Hz, and A = 100 Pa/s and v_{avg} = 0.35 m/s is shown in Figure 6.7 a). The pulsatility is clearly reflected in the heat transfer coefficient, however, the influence of the pulsatility decreases with increasing distance from the inlet, z.

Employing an input pressure profile which is interpolated, P_{int}, from literature [124] using $\frac{1}{c}\frac{\partial P}{\partial t} = P_{int}$ changes the input pulsatile velocity component, u, to reflect the complex blood flow dynamics of the aorta. The simulated heat transfer coefficient at 0.01 m is similar in shape to P_{int}, however, both the shape and the influence of the pulsatile input decrease with increasing distance from the inlet.

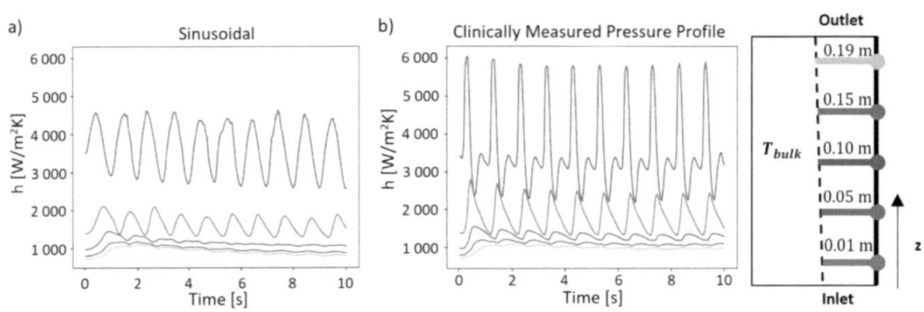

Figure 6.7: Pulsatile investigation of the time dependent determination of the heat transfer coefficient. Open Access figure adapted from [1] published under open access under CC BY license (Creative Commons Attribution 4.0 International license).

Figure 6.8 shows the averaged heat transfer coefficients plotted against the averaged velocities (last 5 seconds of each simulation as shown in the Appendix in Figure F.3) with different parameter variations. In addition to the different color-coded distances from the inlet (z = 0.01 m, 0.05 m, 0.10 m, 0.15 m, and 0.19 m), the frequency f = 1 Hz and 2 Hz and the amplitude A = 50 Pa/s, 100 Pa/s, and 200 Pa/s are varied. Notably, the heat transfer coefficient h, decreases with increasing distance from the inlet in all parameter variations. Further, as the averaged velocity increases, the heat transfer coefficient increases. This tendency is also displayed in Figure 6.6, where the fluid velocity is increased through the tapering of the cylinder.

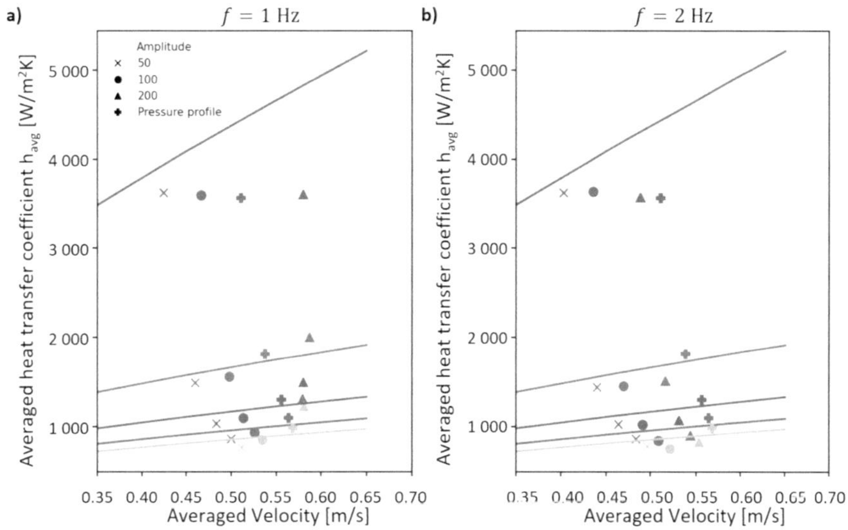

Figure 6.8: Comparison between the heat transfer coefficient for constant velocity (continuous lines) and a parametric sweep of the oscillatory flow (both sinusoidal and interpolated pressure profiles). The color scheme reflects the distance from the cylinder inlet (red: 0.01 m, purple: 0.05 m, blue: 0.10 m, dark green: 0.15 m, and light green: 0.19 m). a) shows the comparison for a frequency of f =1 Hz. b) shows the comparison for a frequency of f =1 Hz. Open Access figure adapted from [1] published under open access under CC BY license (Creative Commons Attribution 4.0 International license).

Figure 6.8 also compares the simulated heat transfer coefficients between the pulsating flow (indicated by symbols) and the constant flow (indicated by continuous lines). The full velocity profile is shown in the Appendix in Figure F.2. For both $f = 1$ Hz and 2 Hz, the flow pulsatility generally decreases the heat transfer coefficient below that for the constant velocity. The corresponding Womersley number $\alpha = 15$ and 22, respectively. However, an increasing average velocity and an increasing distance to the inlet increases the average pulsatile heat transfer coefficient above that of the constant velocity. The heat transfer coefficient with pulsatility for $f = 1$ Hz varies from 25% below to 33% over the heat transfer coefficients for constant flow at the same velocities. For sinusoidal flow with $f = 2$ Hz, the heat transfer coefficient is consistently below that of constant flow with an average of $10 \pm 3\%$. The total overview over the percentage changes for each frequency and amplitude for the sinusoidal pulsatile flow is listed in the Appendix chapter F.1.3.

Opposite to the trend of the heat transfer coefficient for sinusoidal flow, h is increased with pulsatility based on an interpolated pressure profile (except for $z = 0.01$ m). The increase is an average of $7\% \pm 2\%$ for distances between 0.05 m to 0.19 m from the inlet. An increase in the radius is also shown to increase the heat transfer coefficient as shown in the Appendix chapter F.1.4. Despite these variations in the heat transfer coefficient due to different pulsatile flow types and variations, the largest influence remains the distance travelled from the cylinder inlet, z.

In conclusion, the main range of heat transfer coefficients of the simulated pulsatile blood flow is determined from 200 W/m^{2}K to 4800 W/m^{2}K. This range is used in the simulations of the thermodynamic aspect of the blood flow in the next chapter.

6.2 Thermodynamic Aspect of Aortic Blood Flow

The range of relevant heat transfer coefficients in the cylinder modelled after the vascular graft implant is determined between 200 W/m^{2}K and 4800 W/m^{2}K. This range covers the heat transfer coefficient for the constant velocity and oscillating flow from $z = 0.01$ m to 0.19 m. The range of velocities is between 0.35 m/s to 0.65 m/s and covers the relevant physiological velocity range within the implant. The heat transfer coefficient, h, is the parameter which is applied for the thermodynamic simulation of the vascular graft implant (simulation setup described in chapter 3.5.2 and sketched in Figure 3.7). Decoupling the hydrodynamic simulation from the thermodynamic simulation of the material decreases the simulation time significantly. This is possible due to the initial assumption of neglecting material distensibility in the flow simulation. In this chapter, the heat flux flowing through the heat flux sensor (gSKIN XM, resolution of 0.41 W/m^2) is simulated for different heat transfer coefficients, implant materials and thickness, and infectious thermal power and size as listed in Table 6.1. The simulation build is designed to reflect the model system (Figure 6.1) as shown in Figure 3.7 in the theory chapter.

Table 6.1: Parameters swept in the thermodynamic simulation. Values for the materials used in the simulation is listed in Table F.4 in the Appendix.

Thermodynamic simulation parameters for sweep				
Infection thermal power [W/m^3]	Heat transfer coefficient h [W/m^{2}K]	Infection r [mm]	Distance infection and sensor $d_{inf/sensor}$ [mm]	Material
50	200	0.5	0.000	PDMS
200	600	1	0.278	PTFE
1000	1200	2	0.556	ePTFE
2000	2400	4	0.833	Dacron (PETP)
4000	4800	-	1.111	Aortic tissue
10000	-	-	1.389	

In the following parts, two simulations are introduced; first, the determination of the heat transfer efficiency with a sweep of implant material and thickness, and second, a parametric sweep of the aortic implant material/thickness and parametric sweep of the infection (thermal power, size, and distance to sensor) for the determination of the heat flux sensor measurement threshold. The range of heat transfer coefficient between 200 W/m^2K to 4800 W/m^2K is also swept for all of these simulations.

6.2.1 Parametric Sweep of Heat Transfer Efficiency

The parametric sweeps of different implant materials and thicknesses are compared using the heat transfer efficiency [%], which is defined similarly to the heat transfer efficiency in the microfluidic chip (chapter 5.1.2.2). In the simulation, the ratio between the heat flux flowing through the sensor, q_{avg}, and the input heat flux defined at the boundary heat input. The heat flux through the heat flux sensor is defined as the average between the surfaces of the heat flux defined as:

$$q_{avg} = \frac{|\, q_{front\ surface} \,| + |\, q_{back\ surface} \,|}{2} \tag{6.2}$$

$q_{front\ surface}$ and $q_{back\ surface}$ are defined as illustrated in Figure 6.9.

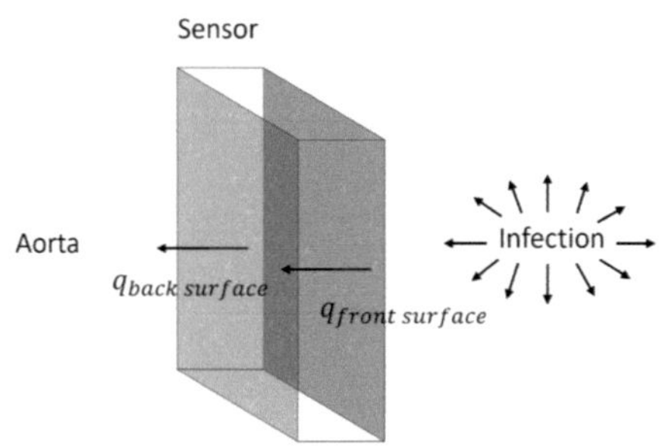

Figure 6.9: Sketch showing the front and back surface of the sensor as labelled in Equation 6.2.

The value of interest is the percentage ratio of total heat flux from the source and what is sensed by the heat flux sensor (heat transfer efficiency [%]). Due to this, the magnitude input heat flux defined at the boundary is irrelevant for the determination of the heat transfer efficiency. In this set of parametric sweeps PDMS, PTFE, ePTFE, Dacron and aortic tissue is compared. The heat transfer efficiency is dependent on the thermal conductivity of the different materials $k_{PMDS} < k_{PTFE} < k_{ePTFE} < k_{aortic\ tissue} < k_{Dacron}$.

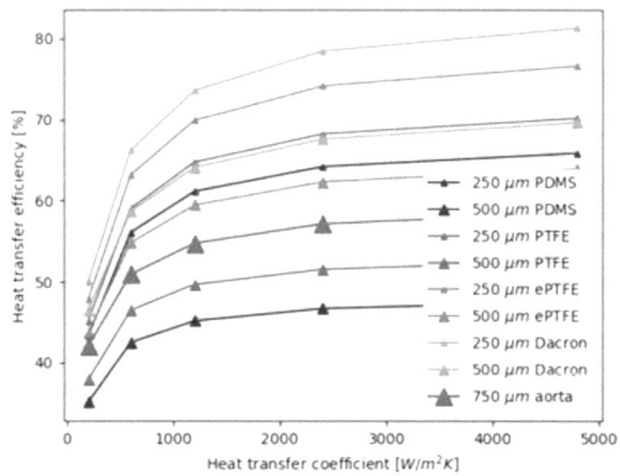

Figure 6.10: Heat transfer efficiency of the aorta in 3D heat transfer simulation with a parametric sweep. Open Access figure adapted from [1] published under open access under CC BY license (Creative Commons Attribution 4.0 International license).

Figure 6.10 shows the heat transfer efficiency for a variation of material thicknesses over the given range of heat transfer coefficients. A vascular graft wall thickness is around 1.0 and 1.5 mm and a sensor thickness of 0.5 mm leaves 250 and 500 µm of vascular graft material, respectively, wrapped around the sensor. The aortic wall thickness is larger than the vascular graft thickness and is normally around 2 mm [145], leaving a hypothetical 750 µm around the sensors. This study is purely for comparison as the heat flux sensor is not intended to be integrated into aortic tissue. Comparing these values, it is clear that the thermal conductivity of the different materials dictate the trend of the heat transfer efficiency if compared at the same thicknesses. Furthermore, increased implant thickness decreases the heat transfer efficiency. The material with the lowest heat transfer efficiency is PDMS, which is used in the simulations in the next sections as a lower bound material. Albeit a thickness of 750 µm a hypothetical sensor embedded in aortic tissue shows larger heat transfer efficiency than both PDMS and PTFE of thickness 500 µm. In regards to the heat transfer efficiency in the case of 250 µm thick PDMS, the heat transfer efficiency increases from 43% to 66% with a heat transfer coefficient varying form 200 to 4800 W/m^{2}K. The increase in heat transfer coefficient decreases with increasing material thickness (all the parameters are listed in the Appendix in chapter F.2.2).

6.2.2 Parametric Sweep Determining the Heat Flux Sensor Measurement Threshold

An input thermal power source can be defined as a volume with a thermal power density with unit [W/m^3]. As already discussed, as illustrated in Figure 5.16, the thermal power density of one *E. coli* bacteria is estimated as 2.35×10^6 W/m^3 and the thermal power density of a bacterial biofilm with 15% active bacteria is estimated as 350000 W/m^3. The calculated biofilm thermal power density is orders of magnitude larger in comparison to literature values for the thermal power density of muscle, 684 W/m^3, and the viscera, 3852 W/m^3 [70]. Throughout this chapter, different input infection thermal power densities, sizes and distances to the sensor are investigated. The goal of the simulation is to determine the threshold at which the simulated heat flux flowing through the heat flux sensor is above its resolution of 0.41 W/m^2. The heat flux flowing through the sensor is defined as shown in equation 6.2.

Figure 6.11 shows the heat flux through the sensor plotted against the range of heat transfer coefficients defined in the fluid dynamic investigations. a), b), c), and d) are the different infectious thermal power densities of 10000 W/m^3, 4000 W/m^3, 2000 W/m^3, and 1000 W/m^3. The infectious source is a cylinder with radius of 4 mm and a thickness of 450 µm. Further, the different colors show increasing distance d between the infection and the sensor. d is calculated as described in the Appendix chapter F.2.7. For an infection placed directly above the heat flux sensor, an infectious thermal power density of 2000 W/m^3 is above the heat flux sensor threshold. A thermal power density of 2000 W/m^3 corresponds to 0.57% of the thermal power density of a biofilm of 350000 W/m^3 (percentage conversion shown in Table F.6). Further simulations for higher thermal power densities are shown in the Appendix in Figure F.5. For a thermal power density of 10000 W/m^3 (2.86% of a biofilm) an infection as far as 5.24 mm away from the heat flux sensor is detectable for a heat transfer coefficient at 600 W/m^{2}K (as indicated with the red circle). Heat transfer coefficients above 2000 W/m^{2}K have little to no influence on the heat flux through the sensor.

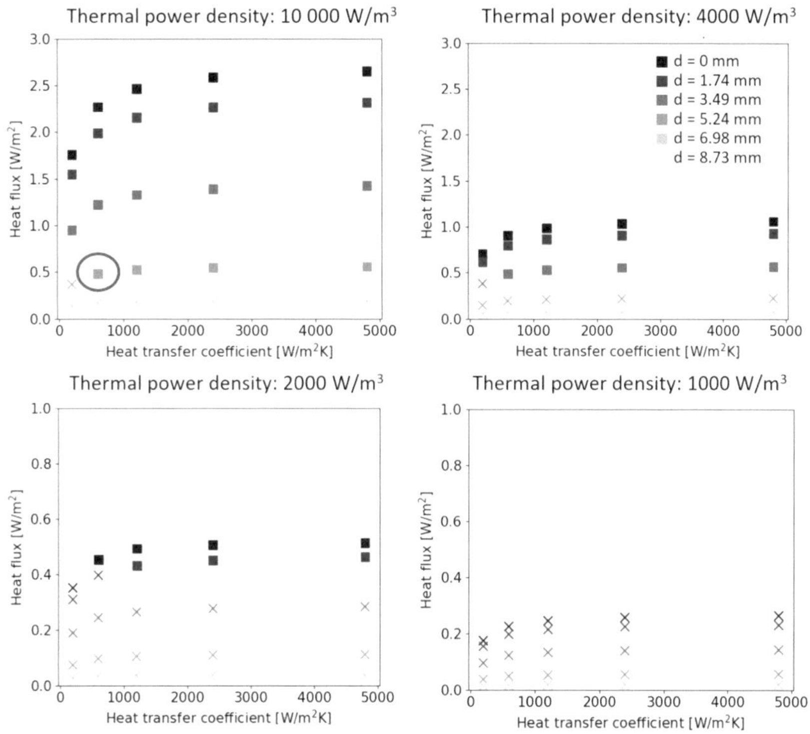

Figure 6.11: Variation in the arc length, *d*, from the center of the sensor to the center of the infection, and the corresponding heat flux measured through the heat flux sensor in the simulation. Data points above the resolution of 0.41 W/m^2 are indicated as squares and blow as crosses. The distance *d* is calculated as described in the Appendix Figure F.8. Open Access figure adapted from [1] published under open access under CC BY license (Creative Commons Attribution 4.0 International license).

Figure 6.12 shows the variation of the same range as in Figure 6.11 for the infectious thermal power density and the heat transfer coefficient. Instead of varying the distance between the sensor and infection, the radius of the infection is varied (constant thickness of 450 µm). For all infectious radii considered (0.5 mm, 1.0 mm, 2.0 mm, and 4.0 mm) the heat flux through the sensor with an infectious thermal power density of 1000 W/m^3 is below the heat flux sensor resolution threshold. For an infectious thermal power density of 2000 W/m^3 an infectious radius of 4.0 mm is above the sensor resolution threshold (as indicated with a red circle in Figure 6.12).

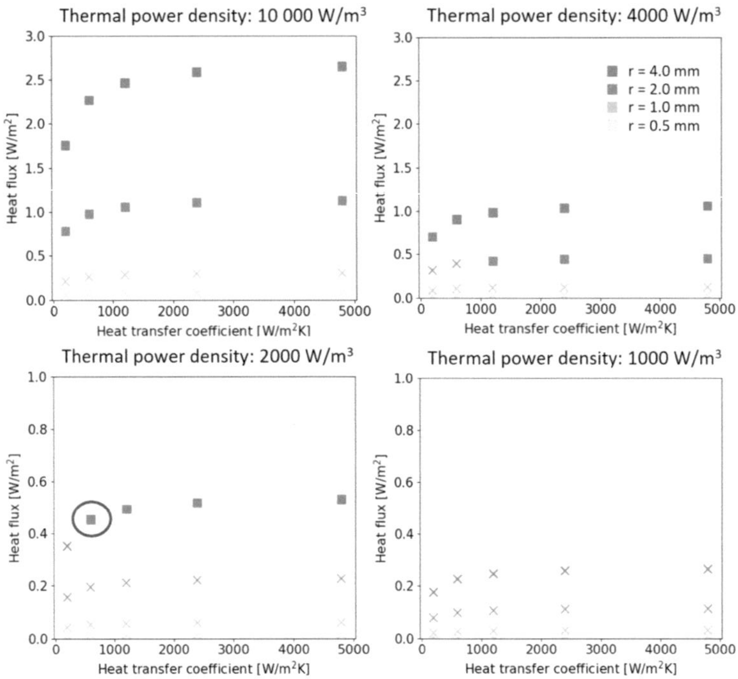

Figure 6.12: Variation in the radius of the thermal power density, r, and the corresponding heat flux measured through the sensor in the simulation. Data points above the resolution of 0.41 W/m^2 are indicated as squares and below as crosses. Open Access figure adapted from [1] published under open access under CC BY license (Creative Commons Attribution 4.0 International license).

The microfluidic chip *in vitro* platform shows experimental detection of a thermal power density of 1707 W/m^3 (gSKIN XM) as discussed in chapter 5.1.2.3 [122]. As shown above, an infection of 4 mm radius with a 450 µm and 2000 W/m^3 is detectable by a heat flux sensor (gSKIN XP). The simulation results and the experimental results align well in regards to the detectability of the thermal power density. In addition, the infectious area volume of the simulation is larger than the microfluidic chip.

As shown in Figure F.7 in the Appendix, the local temperature increases between 0.0025 K and 0.0030 K for an infectious thermal power density of 2000 W/m^3 and and radius 4 mm. Commercial temperature sensors have a resolution limit of 0.1 K, which is two orders of magnitude larger than the temperature change of the system. Thereby, the application of heat flux sensors is preferential for the early detection of infection through bacterial growth on implant surfaces.

6.3 Implementation of Sensor Mesh in the Implant

We envision a system of multiple heat flux sensors around the vascular graft, preferably in a mesh as shown in Figure 6.1. The mesh of sensors allows for differential heat flux measurements as described for the microfluidic chips above (chapter 5.1.2). Intrinsic to this mesh array is the common mode cancellation of surrounding thermal fluctuations as the early stages of infection will only cover a few sensors before spreading. Applying the calculated maximal distance between the sensor and the infection for the early detection of infection determined above (in Figure 6.12), a geometrical mesh covering the implant can be designed to ensure the early detection of infections.

To ensure full coverage infection detection on the implant surface, a 2.86% (infectious thermal power density: 10000 W/m^3) infection detection threshold is defined. For total infection detection coverage, a hexagonal lattice is determined with lattice spacing of $\sqrt{3}\times$ 5.24 mm as shown in the Appendix F.2.7. With this hexagonal lattice, 177 sensors are necessary to cover an implant of 20 cm length and 1 cm radius. Raising the detection threshold to 14.29% (50000 W/m^3), 100 sensors are necessary for full coverage of the implant surface (lattice spacing = $\sqrt{3}$ $\times$6.98 mm). Increasing the lattice spacing also increases the thermal power density threshold at which the infection is detected. These calculations are based on the resolution of a commercially available heat flux sensor (gSKIN XP), which is an available heat flux sensor with the lowest resolution in combination with the smallest size. An implementation of a vascular graft with a mesh of these sensors are yet to be investigated experimentally to determine the system feasibility in regards to infectious growth detection and the mechanical stability of the interface between vascular graft material to sensor.

7 Conclusions and Outlook

The aim of this thesis was to investigate the concept of utilizing heat flux sensors in an aortic vascular graft to detect early infection at the implant. Both the *in vitro* microfluidic chip mimicking the thermal environment of the aortic vascular graft and COMSOL simulations support the feasibility of using heat flux sensors for this concept. Different variations of the experimental chip are shown to reflect the *in vivo* aortic graft environment. In addition, finite element models of different hydrodynamic and thermodynamic conditions allow for a comprehensive understanding of the potential of this concept.

Different thermal power densities and thermal environments at an infection (Figure 2.4) were discussed: the thermal power originating from bacterial growth (in chapter 5.1), the inflammatory cell response (in chapter 5.4), and general blood perfusion in tissue (in chapter 3.2.3). In the *in vitro* microcalorimetric microfluidic chip system built to mimic the thermal environment of the aortic vascular graft, only bacterial growth is measured. Due to experimental limitations, immune cells are not integrated in the microfluidic chip. However, the thermal power originating from a population of THP-1 cells was measured with a microcalorimeter and the thermal power originating from a stimulated immune cell is larger than that of bacteria. Thereby, the microfluidic chip system represents a lower bound of thermal power originating from an infectious site, since only bacteria are considered as the thermal power density at an infectious site. Similarly, finite element models only consider the bacterial population as the thermal power source in addition to considering lower bound thermal transfer materials. The increase in blood perfusion intrinsic in the tissue at an infectious site is not directly relevant for the investigated system, as an increase in heat by blood perfusion increases the temperature at the peripherals rather than in the thorax. This is due to the fact that the temperature in the peripherals is lower than 37°C. Furthermore, the environment considered in this thesis is of smaller scale than the length scale of blood perfused tissue. Thereby, both the experimental and simulation systems present adequate models for the investigations of the feasibility of the early detection of infection.

The microcalorimetric microfluidic chip experiments have been useful to determine the parameter needed to estimate the thermal power density exceeding the heat flux detection limit. This was found to be 1707 W/m^3 (chapter 5.1.2). The thermal power density can be converted to a percentage of the thermal power density of a biofilm (Appendix chapter F.2.3) or an amount of bacteria for a given volume. Therefore the thermal power density limit corresponds to 0.5% of a bacterial biofilm. The region considered to be the infection is the size of the microfluidic channel. The experiment is shown in both thermally stabilized (chapter 5.1.2) and thermally not stabilized environments (chapter 5.3). Furthermore, the material in which the sensor is embedded (PDMS) has a lower thermal conductivity than all vascular graft implant materials (Appendix chapter F.2.1), with better resolution expected for higher thermal conductivity materials (vascular graft materials), as also shown in the finite element modelling chapter. Further, an additional channel on top of the microfluidic chip mimicking the blood flow of the aorta is experimentally shown to increase the heat transfer efficiency by 20% through the sensor (chapter 5.2). These *in vitro* results strongly support that early detection of infection is possible using heat flux sensors.

Besides mimicking the thermal environment of the aortic vascular graft implant, the microcalorimetric microfluidic chip is useful in itself and should be compared to other calorimetric chips presented in literature. Microcalorimeteric, nanocalorimetric, and picocalorimetric chips have been utilized to measure exothermic biological reactions [85–87, 146] in both open [88–90, 147] and closed systems. These systems show great adaptability, however, they remain constrained due to additional parts (heaters, PID controllers, thermally insulating layers). This system does not require additional parts and is more versatile due to the intrinsic PDMS microfluidic chip fabrication properties. This versatility was demonstrated with the double channel microfluidic chip system (chapter 5.2) and with a differential compensation sensing principle to remove the influence of background thermal fluctuations. With this system, both metabolic thermal power originating from bacterial cell growth in addition to exothermic chemical reaction can be measured. The metabolic thermal power originating from a single *E. coli* bacteria can be determined to be between 1.3 to 4.5 pW (in the exponential growth phase), which is consistent with the value reported in literature of 3.5 pW [130].

In this thesis, only the thermal power originating from bacterial growth is measured and the thermal power originating from activated immune cells surrounding the pathogens is not considered in the calorimetric microfluidic chip. However, the thermal power originating from immune cells, as estimated at the end of chapter 5.4, is larger than that originating from *E. coli* bacteria. Thereby, the detection limits determined by the microfluidic chip measurements push the early detection at 0.5% of the bacterial biofilm to even lower amounts of bacteria. The full integration of a microfluidic chip with growing immune cells together with bacteria would require further system optimization and research.

Similar to the microcalorimetric microfluidic chip, the finite element model simulates a thermal power density. This thermal power density is agnostic to its origin which can be considered either bacterial or by immune cells. The range considered in this thesis is so broad that it covers all cases. This thermal power density, similarly to many other parameters, is swept within a large range of values (1000 W/m^3 to 350000 W/m^3), which with the knowledge of the amount of immune cells at an infectious site and the amount of thermal power per cell could also represent any combination of immune cells and bacteria. The fluid dynamic and thermodynamic model of the aortic vascular graft in COMSOL shows results in relation to aortic blood flow (approximated as both sinusoidal flow and interpolated physiological data in chapter 6) not discussed before in literature. In regards to the heat transfer coefficient, h, the distance from the inlet of the simulation, z, shows the largest influence in relation to the other parameters studied as has also been shown in literature [109]. Influences by the sinusoidal pulsatility (A and f) change h from -11% to 33% in relation to the same results for a constant velocity. However, influence of the pulsatility on the flow also decreases with increasing distance from the inlet of the simulation, z. To expand on the simulations studied in this thesis, further simulations in the future should include distensibility, curvatures, and branching of the aortic vascular graft.

The simulated thermal power density threshold for infection detection in relation to the determined heat flux resolution (thermal power density of 2000 W/m^3 above a gSKIN XM sensor) corresponds well to the experimentally determined value (thermal power density 1707 W/m^3 above a gSKIN XP sensor). It is important to note that the PDMS layer is thicker in the simulation (250 µm) than in the *in vitro* experiment (150 µm), which is the reason for the higher infectious thermal power density detection threshold in the simulation. The temperature increase with an infectious thermal power density of 2000 W/m^3 is 3 mK (Appendix chapter F.7), which is orders of magnitudes smaller than standard temperature sensor resolutions (0.1 K). Heat flux sensors give a clear indication of a growing localized thermal power density as opposed to temperature changes which naturally fluctuate over time. The application of differential heat flux sensing as presented experimentally (in chapter 5.1.2.3 and 5.2) allows for significant reduction of the influence of background temperature fluctuations.

In literature, some *in vivo* studies of (mostly orthopaedic) implants seeded with different bacterial concentrations have been presented [139–141]. An infection rate of 50% is shown to be achieved with a colony forming unit (CFU, a measure of viable cells in a bacterial population) of between 2.0×10^6 and 6.2×10^6 cells for *Staphyloccocus aureus* infected on dynamic compression plates fixed to the tibiae [139]. This infection rate is defined as the detected presence of bacteria at the implant site after 28 days [139]. In comparison, the experimentally determined limit of detection of bacterial population for the microfluidic chip is at 9.2×10^5 *E. coli* bacteria (as shown in Table 5.4). In these systems the concentration of bacteria at seeding and at explantation are gathered, no continuous monitoring of the bacterial concentration at the implant is measured. In comparison, the simulated heat flux threshold is at 1.3×10^5 cells, also below the CFU which is used for infection seeding on the implant models *in vivo*.

Upon implantation, the direct injury to the tissue starts the immediate acute inflammation within minutes and lasts from hours to days [148]. At this stage even without bacteria, neutrophils are active at the implant surface trying to phagotycise and thus increasing their metabolism and heat flow. Chronic infections occur when the monocytes and macrophages are present at the implant [148]. If there are pathogens present, the inflammation/infection could last weeks to months, during which the macrophages persist and continuously battle the pathogens. Early infection occurs between 2 to 4 days to within 3 months of implantation [149, 150]. Delayed infection occurs between 3 and 24 months, and late infection occurs after 24 months [150]. In relation to the measurement of heat flux related purely to infection and not inflammation, a later onset infectious growth would be easier to resolve. However, the increase in heat flow due to bacterial growth is expected to significantly increase the local heat flow. Therefore, a possibility of differential heat flux detection of bacterial growth during the acute inflammation phase could be possible. However, it would have to be investigated in an *in vitro* model in future work.

In chapter 6.3, the possibility of early detection of infection is theoretically investigated with an unknown location of the onset infection, but it could happen at any point along the graft. To do this, a hexagonal mesh design is shown with the theoretical possibility for full coverage early detection of infection on the aortic vascular implant. Further, this system allows for differential sensing of infection between the heat flux sensors. The inherent advantages of the heat flux sensing system is applied allowing for higher resolution as opposed to using temperature sensing (chapter 6.2.2). As opposed to thermoelectric-effect based heat flux measurements (∂T between top and bottom side of heat flux sensor), temperature measurements (T_{abs}) do not allow for distinguishing general temperature increase vs. localized increase in heat flow. This system is yet to be investigated *in vivo* which requires further system development and integration. The intrinsic influence of vascular graft curvature and material differences at the interface between the sensors and the vascular graft implant material are aspects which need to be investigated and optimized.

Overall, the presented experimental and simulation results show that heat flux sensors have great potential for the early detection of infection on implant surfaces. These results show the possibility of infection detection on aortic vascular graft implants, mainly for the additional benefit of the intrinsic heat drain (blood flow). However, as these results show the possibility of bacterial growth detection at relatively low amounts of bacteria, the heat flux detection system could be expanded to other implant surfaces than just aortic vascular grafts which have an intrinsic heat drain. The development of a calorimetric microfluidic chip system measuring the interaction between immune cells and bacteria would give further insight into the localized heat flow at any implant surface. In addition, explorations of simulation systems integrating aortic branching, curvature, and distensibility allows for further insight in the *in vivo* system. Similarly to the investigations in literature monitoring the initial bacterial concentrations for the assessment of implant infections [139–141], an implant with heat flux sensors should be investigated *in vivo* in an animal model. Any further investigations on the use of heat flux sensors for the early detection of infection on aortic vascular grafts are greatly supported by the experiments and simulations presented in this thesis.

A Appendix - Fluid Dynamics Theory

A.1 Fluid Dynamic Parameters of Water and Blood

Throughout this thesis, two different liquids are of interest; water and blood. In this thesis we use blood and water. The general blood and water parameters are shown in Table A.1. The specific parameters for the aorta at different velocities is shown in Table A.2. The flow environment within the microfluidic chip for 320 µm×12 mm×12 mm is shown in Table A.2.

Table A.1: The fluid dynamics parameters of blood and water.

Liquid properties of blood and water at 37°C						
Fluid	Density ρ [kg/m^3]	Dynamic viscosity μ_{dyn} [Pa s]	Specific heat c_p [J/kgK]	Thermal conductivity k [W/mK]	Kinematic viscosity v [m^2/s]	Prandtl number Pr
Blood	1060	0.00278	3617	0.54	2.62 $\times 10^{-6}$	18.6
Water	997	0.00069	4184	0.61	6.94 $\times 10^{-7}$	4.7

Table A.2: Table of different hydrodynamic values of blood relevant for the physiological conditions of an aorta of radius of 1 cm.

Fluid dynamic parameters			
Velocity [m/s]	Reynolds number Re	Hydrodynamic entrance length L_h [m]	Thermodynamic entrance length L_{th} [m]
0.11	838	0.84	15.62
0.35	2669	2.66	49.80
0.65	4957	5.00	92.30

The characteristic length scale L_{ch} is normally defined as $L_{ch} = 2hb/(h+b)$ for rectangular channels. However, Muzychka and Yovanovich [120] have defined a more fitting characteristic length scale for non-circular ducts $L_{ch} = \sqrt{hb}$. For cylinders the characteristic length scale is $L_{ch} = D$.

Table A.3: Table of the calculated flow parameters in water in the microfluidic chip with dimensions 320 µm×12 mm×12 mm and $L_{ch} = 2hb/(h+b)$.

Microfluidic flow environment and flow rates with $L_{ch} = 2hb/(h+b)$			
Flow rate $[m^3/s]$	Reynolds number Re	Hydrodynamic entry length L_h [µm]	Thermal entry length L_{th} [µm]
1.875×10^{-10}	0.050	1.7	8.2
3.333×10^{-10}	0.08	3.1	14.5
4.271×10^{-10}	0.112	3.9	18.6
1.125×10^{-9}	0.300	10.0	48.9
1.646×10^{-9}	0.433	15.1	71.6

Table A.4: Table of the calculated flow parameters in water in the microfluidic chip with dimensions 320 µm×12 mm×12 mm and $L_{ch} = \sqrt{\text{hb}}$.

Microfluidic flow environment and flow rates for $L_{ch} = \sqrt{\text{hb}}$			
Flow rate [m^3/s]	Reynolds number Re	Hydrodynamic entry length L_h [µm]	Thermal entry length L_{th} [µm]
1.875×10^{-10}	0.014	4.8	22.9
3.333×10^{-10}	0.246	8.6	40.7
4.271×10^{-10}	0.315	11.0	52.1
1.125×10^{-9}	0.830	29.0	137.2
1.646×10^{-9}	1.214	42.4	200.8

A.2 Viscous Dissipation

At small length scales such as in microchannels, different effects dominate and might even change different parameters. The effect of viscous dissipation has to be considered, especially when the localized thermal output is measured. Xu et al. have developed an evaluation of the viscous dissipation of liquid flow in microchannels [151]. They define a viscous number Vi:

$$Vi = \frac{\mu_{dyn} \mathrm{v}_{avg}^2 L_{ch}}{\rho \mathrm{v}_{avg} c_p T_{ref} D^2} \tag{A.1}$$

with μ_{dyn} as the viscosity, v_{avg} as the average velocity, $L_{ch} = \sqrt{\mathrm{hb}}$ as the characteristic length, ρ as the density, c_p as the heat capacitance, T_{ref} as the reference temperature (chosen as 1 K) and D as the diameter. Further they describe the temperature difference upon viscous dissipation as:

$$\partial T = \frac{93.419 ViPr^{-0.1}}{5.2086 + ViPr^{-0.1}} \tag{A.2}$$

They describe the indication of the significance of viscous dissipation through $ViPr^{-0.1}$ [151]:

$$ViPr^{-0.1} \leq 0.056 \tag{A.3}$$

This indicates no viscous dissipation effect.

$$ViPr^{-0.1} \geq 0.056 \tag{A.4}$$

This indicates significant viscous dissipation effects.

Table A.5: Viscous dissipation for the microfluidic chip dimensions used in this thesis.

System	Flow rate [µL/min]	Dimensions	v_{avg} [m/s]	$ViPr^{-0.1}$
Microfluidic Chip [122]	98.75	320 µm × 12 mm × 12 mm	4.3×10^{-4}	1.9×10^{-12}
Microfluidic Chip [122]	11.25	320 µm × 12 mm × 12 mm	4.9×10^{-5}	2.1×10^{-13}
Microfluidic Chip [126]	98.75	170 µm × 10 mm × 10 mm	9.7×10^{-4}	2.8×10^{-12}
Microfluidic Chip [123]	200.00	1 mm × 12 mm × 12 mm	2.8×10^{-4}	2.1×10^{-12}

In all of the systems in Table A.5, $ViPr^{-0.1} < 0.056$. Thereby, there are no significant viscous dissipation effects in the microfluidic chips used in this thesis.

A.3 Different Cylinder Flow Regions with Varying Radii

Figure A.1 shows the different flow regimes in a cylinder with varying radii from 0.004 m to 0.014 m.

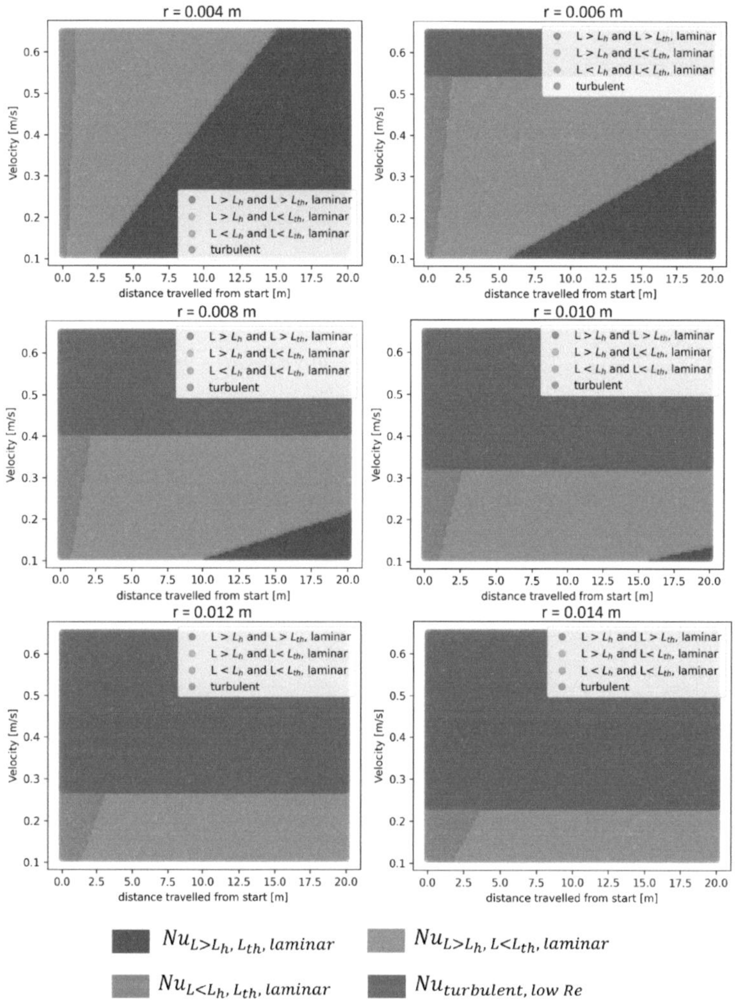

Figure A.1: Flow regimes with different cylinder radius. The range of radius chosen are in the physiological regime for children and adults. The length of the distance from the inlet is increased to show the turbulent flow regime.Open Access figure adapted from [1] published under open access under CC BY license (Creative Commons Attribution 4.0 International license)

A.4 Determination of the Heat Transfer Coefficient in Laminar Flow

A cylindrical system is considered with constant heat flux at the wall (as opposed to constant temperature) with q_s=constant, with hydrodynamically fully developed flow environment $\frac{\partial \mathrm{v}(r,z)}{\partial z} = 0$ ($\mathrm{v} = \mathrm{v}(r)$), and with thermodynamically fully developed environment $\frac{\partial}{\partial z}[\frac{T_s(x)-T(r,z)}{T_s-T_m(z)}]$ (in combination with constant heat flux this becomes $\frac{dT}{dz}$=constant). Due to the constant radius and the laminar flow, many different simplifications of the equations can be made. This section is a condensed text based on a textbook called Heat transfer; A Practical Approach [118].

A.4.1 Average Velocity v_m

Applying the principle of conservation of mass, the average velocity can be determined. The mass conservation in a cylinder can be expressed as following:

$$\dot{m} = \rho \mathrm{v}_m A_c = \int_{A_c} \rho \mathrm{v}(r,z)\, dA_c \tag{A.5}$$

with $\dot{m}$ as the mass flow rate and $\mathrm{v}(r,z)$ as the velocity profile in the cylinder. The average velocity v_m can be expressed by rearranging the expression and expressing the surface integral with a radial integral (possible due to inherent symmetry of the cylinder):

$$\mathrm{v}_m = \frac{\int_{A_c} \rho \mathrm{v}(r,z)\, dA_c}{\rho A_c} = \frac{\int_0^R \mathrm{v}(r) 2\pi r\, dr}{\pi R^2} = \frac{2}{R^2} \int_0^R \mathrm{v}(r) r\, dr \tag{A.6}$$

Assuming fully developed flow the radial dependent velocity of the axial component of velocity is expressed by A.7, which is the well-known Hagen-Poiseuille equation:

$$\mathrm{v}(r) = -\frac{R^2}{4\mu}(\frac{dP}{dz})(1 - \frac{r^2}{R^2}) \tag{A.7}$$

Inserting this equation in Equation A.6 yields:

$$\mathrm{v}_m = \frac{2}{R^2}\int_0^R -\frac{R^2}{4\mu}(\frac{dP}{dz})(1-\frac{r^2}{R^2})r\,dr = -\frac{R^2}{8\mu}\frac{dP}{dz} \tag{A.8}$$

Which can be integrated in the expression of $\mathrm{v}(r)$ (Equation A.7):

$$\mathrm{v}(r) = -2\frac{R^2}{8\mu}(\frac{dP}{dz})(1-\frac{r^2}{R^2}) = 2\mathrm{v}_m(1-\frac{r^2}{R^2}) \tag{A.9}$$

A.4.2 Average Temperature T$_m$

Applying the principle of energy conservation and Equation A.5 the energy flow of the fluid, $\dot{E}_{fluid}$, can be expressed as:

$$\dot{E}_{fluid} = \dot{m}c_pT_m = \int_{A_c} \rho c_p T(r)\mathrm{v}(r)\,dA_c \tag{A.10}$$

With c_p as the specific heat capacity of the fluid in [J/(kg K)].

By rearranging we can express the average temperature T_m:

$$T_m = \frac{\int_{A_c}\rho c_p T\mathrm{v}\,dA_c}{\dot{m}c_p} = \frac{\int_r c_pT\rho\mathrm{v}(2\pi r dr)}{\rho\mathrm{v}_m(\pi R^2)c_p} = \frac{2}{\mathrm{v}_m R^2}\int_0^R T(r)\mathrm{v}(r)r\,dr \tag{A.11}$$

The energy balance between the outside of the cylinder to within can be expressed through this equation:

$$\dot{m}c_pT_z - \dot{m}c_pT_{z+dz} + \dot{Q}_r - \dot{Q}_{r+dr} = 0 \tag{A.12}$$

where $\dot{m}c_pT_z$ and $\dot{m}c_pT_{z+dz}$ are the energy flows of the fluid and $\dot{Q}_r$ and $\dot{Q}_{r+dr}$ are the rate of heat transfer in the radial direction. Rearranging and applying Equation A.5 in addition to applying $dA_c = 2\pi r dr$ and dividing through $2\pi r dr dx$ one gets:

$$\rho c_p\mathrm{v}(r)\frac{T_{z+dz}-T_z}{dz} = -\frac{1}{2\pi r dz}\frac{\dot{Q}_{r+dr}-\dot{Q}_r}{dr} \tag{A.13}$$

which can again be rearranged to:

$$\mathrm{v}(r)\frac{\partial T}{\partial z} = -\frac{1}{2\rho c_p \pi r dz}\frac{\partial \dot{Q}}{\partial r} \tag{A.14}$$

with the definition of heat flow as $\dot{Q} = dA_r q = 2\pi r dz q = -2\pi r dz k\frac{\partial T}{\partial r}$ (A_r is the area considered for the radial heat flux) this can be expressed as:

$$\mathrm{v}(r)\frac{\partial T}{\partial z} = \frac{k}{r\rho c_p}\frac{\partial}{\partial r}(r\frac{\partial T}{\partial r}) \tag{A.15}$$

This equation states that the net rate energy transfer by mass flow in z direction is equal to net rate heat conduction in r direction.

A.4.3 Determining the Heat Transfer Coefficient

The temperature differential along the flow direction for constant heat flux (q_s) in the fully developed flow region is expressed as:

$$\frac{\partial T}{\partial z} = \frac{dT_s}{dz} = \frac{dT_m}{dz} = \frac{2q_s}{\rho \mathrm{v}_m c_p R} = constant \tag{A.16}$$

Applying this equation and Equation A.9 in Equation A.15 one gets:

$$\mathrm{v}(r)(\frac{2q_s}{\rho \mathrm{v}_m c_p R}) = \frac{k}{r\rho c_p}\frac{d}{dr}(r\frac{\partial T}{\partial r}) \tag{A.17}$$

$$2\mathrm{v}_m(1-\frac{r^2}{R^2})(\frac{2q_s}{\rho \mathrm{v}_m c_p R}) = \frac{k}{r\rho c_p}\frac{d}{dr}(r\frac{\partial T}{\partial r}) \tag{A.18}$$

By solving the differential equation and rearranging this yields a general solution of $T = \frac{q_s}{kR}(r^2 - \frac{r^2}{4R^2}) + C_1 r + C_2$, applying $\frac{\partial T(r=0)}{\partial z} = 0$, $T(r=R) = T_s$ one gets:

$$T(r) = T_s - \frac{q_s R}{k}(\frac{3}{4} - \frac{r^2}{R^2} + \frac{r^4}{4R^4}) \tag{A.19}$$

Inserting this equation in equation A.11 one gets:

$$T_m = \frac{2}{\mathrm{v}_m R^2}\int_0^R (T_s - \frac{q_s R}{k}(\frac{3}{4} - \frac{r^2}{R^2} + \frac{r^4}{4R^4}))(2\mathrm{v}_m(1-\frac{r^2}{R^2}))r\,dr \tag{A.20}$$

Solving this and integrating gives:

$$T_m = T_s - \frac{11}{24}\frac{q_s R}{k} \tag{A.21}$$

Applying $q_s = h(T_s - T_m)$ then finally gives the velocity-independent expression of the heat transfer coefficient in the laminar fully developed flow condition:

$$h = \frac{24}{11}\frac{k}{R} = \frac{48}{11}\frac{k}{D} = 4.36\frac{k}{D} \tag{A.22}$$

This is specifically for the case of constant heat flux, q_s. This equation shows the heat transfer coefficient in the laminar thermally and fluid dynamically fully developed flow region. For constant temperature T_s, $h = 3.66\frac{k}{D}$.

A.5 COMSOL Partial Differential Equation

The axial component of the momentum equation can be written as follows:

$$\rho\frac{\partial}{\partial t}v = -\frac{\partial}{\partial z}P + \mu\frac{1}{r}\frac{\partial}{\partial r}(r\frac{\partial}{\partial r}v) \tag{A.23}$$

Since the pulsatile velocity u is investigated using the partial differential equation v is substituted by u.

$$\rho\frac{\partial}{\partial t}\mathrm{u} = -\frac{\partial}{\partial z}P + \mu\frac{1}{r}\frac{\partial}{\partial r}(r\frac{\partial}{\partial r}\mathrm{u}) \tag{A.24}$$

The equation is rearranged as follows:

$$-\frac{\partial}{\partial z}P = \rho\frac{\partial}{\partial t}\mathrm{u} - \mu\frac{1}{r}\frac{\partial}{\partial r}(r\frac{\partial}{\partial r}\mathrm{u}) \tag{A.25}$$

$\frac{\partial}{\partial z}P$ can be expressed as $\frac{1}{c}\frac{\partial}{\partial t}P$ and by applying the product rule it yields:

$$-\frac{1}{c}\frac{\partial}{\partial t}P = \rho\frac{\partial}{\partial t}\mathrm{u} - \mu(\frac{\partial^2\mathrm{u}}{\partial r^2} + \frac{1}{r}\frac{\partial\mathrm{u}}{\partial r}) \tag{A.26}$$

To solve for the pulsatile element of the flow, u, in the time dependent study, a general partial differential equation (PDE) given in COMSOL is used:

$$e_a\frac{\partial^2\mathrm{u}}{\partial t^2} + d_a\frac{\partial\mathrm{u}}{\partial t} + \nabla(-c\nabla\mathrm{u} - \alpha\mathrm{u} + \gamma) + \beta\nabla\mathrm{u} + a\mathrm{u} = f \tag{A.27}$$

$e_a, d_a, c, \alpha, \gamma, \beta$ and f are different placeholder coefficients that can either be filled with values or with 0 to construct the desired PDE. By replacing the different coefficients with the values relevant to the partial differential equation which describes the Womersley flow, the pulsatile velocity is introduced in the flow simulation. In this equation $e_a = 0$ is the absorption coefficient, $d_a = \rho_b$ is the damping coefficient, $c = \mu_b$ is the diffusion coefficient, $\alpha = 0$ is the conservative flux convection coefficient, $\gamma = 0$ is the conservative flux source, $\beta = -\mu_b/r$ is the convection coefficient, $a = 0$ is the adsorption coefficient, and $f = -\frac{1}{c}\frac{\partial P}{\partial t}$ is the source term (with P using a *cos* function). Thereby, the pulsatile flow equation is expressed as:

$$\rho\frac{\partial\mathrm{u}}{\partial t} + \nabla(-\mu\nabla\mathrm{u}) - \mu/r\nabla\mathrm{u} = -\frac{1}{c}\frac{\partial P}{\partial t} \tag{A.28}$$

B Appendix - Lumped Element Model Parameters

The following part of the Appendix is based on two previously published papers in literature [122, 126].

B.1 One Level Microfluidic Chip

The different material parameters used for the one level microfluidic chip are shown in Table B.1. The different geometric parameters used for this microfluidic chip is shown in Table B.2. The extracted lumped element model parameters are shown in Table B.3.

Table B.1: LEM material parameters for one level microfluidic chip.

Material	k [W/mK]	Density ρ [kg/m^3]	Specific heat c_p [J/kgK]	h_{water} [W/m^{2}K]
PDMS	0.16	965	1460	-
TEG (Bi_2Te_3)	1.2	7700	154	-
PMMA	0.2	1185	1466	-
Glass	1.38	2200	770	-
Water	-	-	-	9008.4

The simulation values for the TEG is assumed to be the material properties of Bi_2Te_3 for simplicity.

Table B.2: LEM geometric parameters for one level microfluidic chip.

Area	10^{-4} [m^2]	Length	10^{-6} [m]
A_{PDMS1}	1.44	l_{PDMS1}	150
A_{PDMS2}	0.44	l_{PDMS2}	500
A_{PDMS3}	1.44	l_{PDMS3}	150
A_{glass}	1.44	l_{glass}	170
A_{TEG}	1.00	l_{TEG}	500
A_{PMMA}	1.44	l_{PMMA}	1000
A_{air}	1.44	l_{air}	100

Table B.3: LEM parameters for one level microfluidic chip.

Material layer	Thermal resistance R	Thermal capacitance C	Time constant τ [s]
PDMS 1	6.5	3.04×10^{-2}	0.20
TEG	4.2	5.93×10^{-2}	0.25
PDMS 2	71	3.10×10^{-2}	2.20
PDMS 3	6.5	3.04×10^{-2}	0.20
Glass	0.9	4.55×10^{-2}	0.04
PMMA	34.7	2.50×10^{-1}	8.68

B.2 Double Channel Microfluidic Chip

The different values from the one level lumped element model is shown in Table B.4. The material parameters are used as shown in Table B.1.

Table B.4: Table of values used for the double channel lumped element model.

Overview over Lumped Element Model Parameters Double Channel							
Parameter	PDMS	TEG	$Water_{top}$	$Water_{bottom}$	Air	PMMA	Glass
A [mm^2]	144 or 44	100	144	144	144	144	144
l [µm]	150 or 500	500	-		200	1000	170
h [W/m^2K]	-	-	4750 or 9009		112	-	-
k [W/mK]	0.16	1.20	-		0.03	0.02	1.28

C Appendix - Datasheet

C.1 greenTEG gSKIN Heat Flux

gSKIN® Heat Flux Sensors for R&D

- Ultra-high resolution of thermal energies and temperature differences
- Low invasiveness & thickness
- Versions with connectors compatible with all gSKIN® DLOG data loggers
- All sensors with conductive heat flux calibration cohering to ISO 8301

gSKIN®-XM   gSKIN®-XP   gSKIN®-XI   gSKIN®-XO   gSKIN®-XO-Temp

Product Name	gSKIN® XM 26 9C	gSKIN® XM 27 9C	gSKIN® XP 26 9C	gSKIN® XP 27 9C	gSKIN® XI 26 9C	gSKIN® XI 27 9C	gSKIN® XO 67 7C	gSKIN® XO-66 7C -Temp
Article Number	A-044336	A-044339	A-044573	A-044577	A-044628	A-044630	A-044717	A-044728
Detector Type	Thermoelectric		Thermoelectric		Thermoelectric		Thermoelectric	Thermoelectric / NTC
Surface Material (Sensing Area)	Anodized Aluminum		Anodized Aluminum		Anodized Aluminum		Polyamide	Polyamide
Sensing Dimensions (a x b x d) [mm x mm x mm]	4.4 x 4.4 x 0.5		10.0 x 10.0 x 0.5		18.0 x 18.0 x 0.5		30.0 x 30.0 x 3.3	30.0 x 30.0 x 3.3
Heat Flux Range Min / Max [kW/m²]	-150 / 150		-150 / 150		-150 / 150		-15/ 15	-15/ 15
Noise Equivalent Heat Flux* [W/m²] / absolute [µW]	0.34 / 6.6		0.05 / 5.0		0.02 / 4.6		0.07 / 64.3	0.07 / 64.3
Heat Flux Resolution [W/m²] / absolute [µW] with gSKIN® DLOG*	0.41 / 7.9		0.06 / 6.1		0.02 / 5.7		0.09 / 78.4	0.09 / 78.4
Min/Avg* Sensitivity (S) [µV/(W/m²)]	1.5 / 4.0		10.0 / 20.0		35.0 / 65.0		7.0 / 13.0	7.0 / 13.0
Temperature Dependence* of S [%/°C]	0.25		0.25		0.25		0.25	0.25
Response Time* (0→95%) [s]	0.7		0.7		0.7		n/a	n/a
Electrical Resistance [Ohm]	<20		<150		<400		<150	<150
Thermal Conductivity [W/(m*K)]	~1.1		~1.2		~1.3		~0.4	~0.4
Max. Compressive Force when clamped [kgf]	< 2		<10		<32		>32 (not specified)	>32 (not specified)
Operating Temperature Range Min/Max [°C]	-50 / 150		-50 / 150		-50 / 150		-50 / 80	-50 / 80
Calibration Temperature Range Min/Max [°C]	-30 / 70		-30 / 70		-30 / 70		-30 / 70	-30 / 70
Calibration Accuracy [±%]	3		3		3		3	3
Homogeneity* [±%]	1		1		1		1	1
Linearity with Power [±%]	1		1		1		1	1
Accuracy Temperature Measurement [°C]	n/a		n/a		n/a		n/a	+/- 0.2 (0-30 °C), +/- 0.4 (-20-60 °C)
Resolution Temperature Measurement [°C]	n/a		n/a		n/a		n/a	n/a
Flexprint Length (f) [cm]	5		5		5		n/a	n/a
Cable Length (c) [cm] (Connector: no/yes)	100 (no)	100 (yes)	100 (no)	100 (yes)	100 (no)	100 (yes)	100 (yes)	100 (no)

* Experimentally evaluated values under optimal steady state conditions
* Guaranteed minimum heat flux resolution using the gSKIN® DLOG-4219 (not applicable for XO-Temp)
* Only a reference point
* The sensitivity increases (decreases) as the temperature goes above (below) 22.5 °C
* Refers to the heat flux measurement
* Conductive heat flux calibration cohering to the ISO8301 standard with mean temperature of 22.5 °C
* Position dependent signal change across sensing area

greenTEG AG   Hofwisenstr. 50A   8153 Rumlang, Switzerland   T: +41 44 515 09 15   info@greenTEG.com   greenTEG.com

Figure C.1: Copy of datasheet of the gSKIN heat flux sensors used in this thesis.

With the given heat flux resolution of gSKIN XM of 0.41 W/m^2 and of gSKIN XP of 0.06 W/m^2 in addition to their respective thermal conductivities of k = - 1.1 W/(m^{2}K) and k = - 1.2 W/(m^{2}K) and the sensor thickness of 0.5 mm, one can calculate the needed corresponding temperature increase using this formula:

$$q = -k\frac{dT}{dx} \tag{C.1}$$

$$dT = -\frac{qdx}{k} \tag{C.2}$$

This yields a $dT = 0.0002$ K for gSKIN XM and a $dT = 0.00003$ K for gSKIN XP.

D Appendix - Experimental System

D.1 Pictures of thermally stabilized and thermally not stabilized systems

Pictures of both the thermally not stabilized and thermally stabilized systems are shown in Figure D.1 (thermally not stabilized) and Figure D.1 (thermally stabilized).

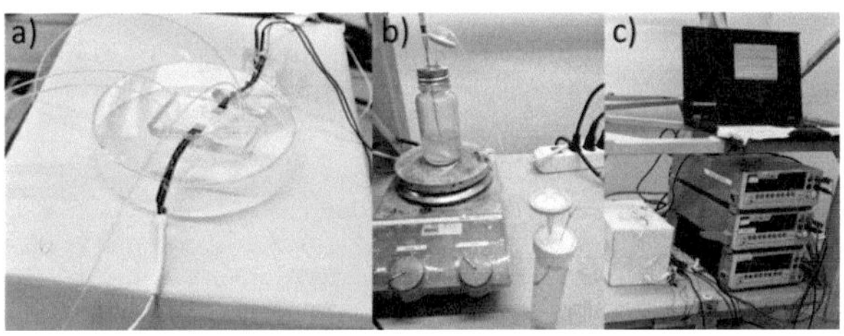

Figure D.1: a) shows a close-up picture of the microfluidic chip. b) shows the LB media with bacteria and sterile LB media. c) shows the measurement setup. Open Access figure from [125] published under open access under CC BY license (Creative Commons Attribution 4.0 International license).

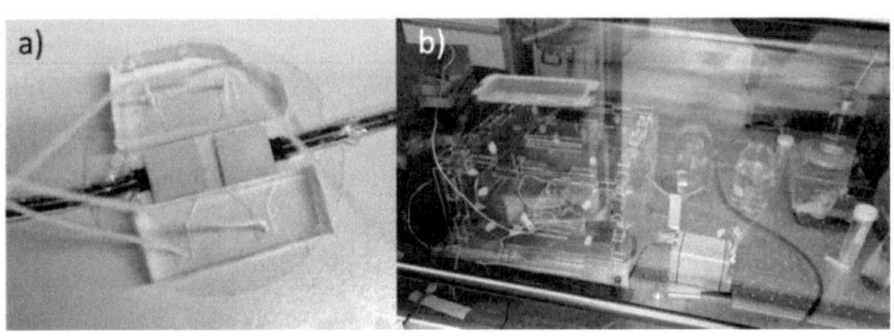

Figure D.2: a) shows a close-up picture of the microfluidic chip. b) shows a picture of the system inside the incubator. Open Access figure from [125] published under open access under CC BY license (Creative Commons Attribution 4.0 International license).

E Appendix - Microfluidic Chip Experiment

The following part in the Appendix is based on previously published results in literature [122].

E.1 Comparison Between Inlet and Outlet Bacterial Growth

The optical density in the bottle and at the end of the outlet of the microfluidic chip is measured in parallel as shown in Figure E.1.

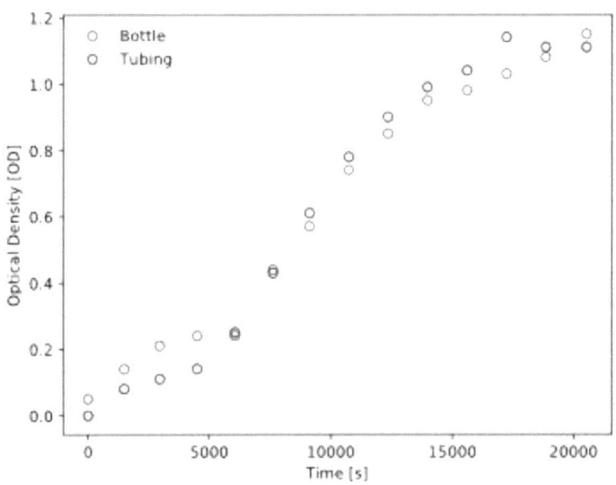

Figure E.1: Comparison between OD measurement before and after travelling through tubing and microfluidic chip. Open Access figure from [122] published under open access under CC BY license (Creative Commons Attribution 4.0 International license).

The purpose of this experiment is the comparison between the bacterial growth curves of *E. coli* in an aerated environment and after having travelled through the tubing and the microfluidic chip. A sample of the media containing bacteria is collected every 25 min and placed on ice. The flow rate of the peristaltic pump is 25.6 µL/min and the distance travelled is 100 cm of PTFE in addition to 30 cm of peristaltic tubing. Figure E.1 is corrected for the travel time in the tubing. In the first 6000 s and towards the end in the saturation phase, a slight difference between the two systems is visible. However, a good overlap is visible during the exponential growth phase. An explanation for the initial lag could be that the bacteria stick to the tubing wall. Towards the end, an explanation could be that the bacteria that initially stick to the wall release and get collected.

E.2 Determination of the OD to bacteria/mL Conversion

Agar plates with LB are prepared and samples are plated every 30 min from a bacterial growth curve. In parallel, the optical density is measured of the bacterial growth curve. For the plating a 10-fold dilution is made using sterile PBS (7-fold). The amount of bacteria is counted using ImageJ. Both the measured optical density (x-axis) and the counted amount of bacteria (y-axis) are shown in Figure E.2. The consequent fitting, which is also shown, allows for the extraction of a slope which corresponds to the conversion value: $1.08 \times 10^9 = OD_{ratio}$. This conversion value changes the values from optical density to bacteria/mL.

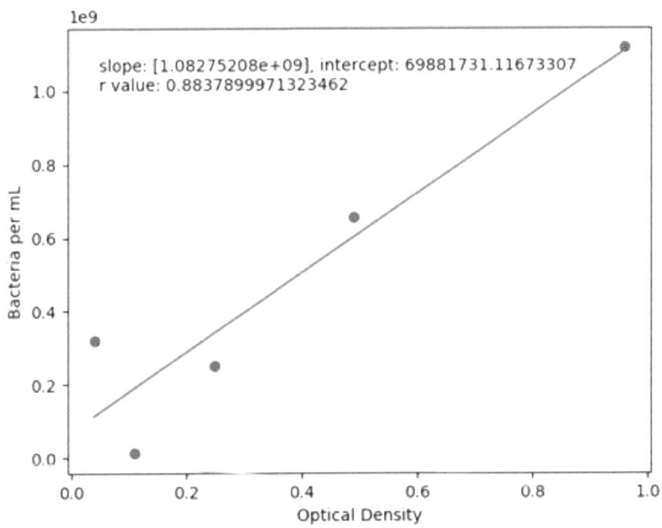

Figure E.2: Determination of the conversion factor between OD and bacteria/mL. Open Access figure from [122] published under open access under CC BY license (Creative Commons Attribution 4.0 International license).

E.3 Error Propagation

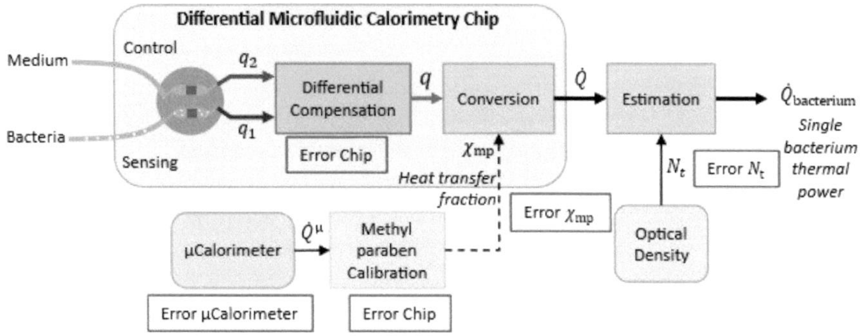

Figure E.3: Schematic overview of error propagation. Open Access figure from [122] published under open access under CC BY license (Creative Commons Attribution 4.0 International license).

The general expression for the exponential fit is as follows:

$$y = aexp(-bx) - c \tag{E.1}$$

The values are scaled to 46 µL by multiplying the ratio between 46 µL and 20 µL to the values of a. Applying the error propagation to $X_{mp} = \dot{Q}_{m0}/\dot{Q}_0^{\mu}$ gives the following expression:

$$X_{mp} = Xmp\sqrt{(\frac{\partial \dot{Q}_0^{\mu}}{\dot{Q}_0^{\mu}})^2 + (\frac{\partial \dot{Q}_{m0}}{\dot{Q}_{m0}})^2} \tag{E.2}$$

Then the exponential growth phase of the bacteria is fit to an exponential function (both OD measurement and the heat flow measured):

$$ln(y) = at + b \tag{E.3}$$

The respective values and errors are determined by using the standard deviation of the residual as the error as shown in E.1. The single bacteria thermal power is defined as:

$$\dot{Q}_{bacteria} = \frac{\dot{Q}/X_{mp}}{N_t OD_{ratio} V_{ratio}} \tag{E.4}$$

The error applied to this expression is calculated as:

$$\partial\dot{Q}_{bacteria} = \dot{Q}_{bacteria}\sqrt{(\frac{\partial X_{mp}}{X_{mp}})^2 + (\frac{\partial\dot{Q}}{\dot{Q}})^2 + (\frac{\partial N_t}{N_t})^2} \tag{E.5}$$

Where only the errors of $\dot{Q}$ (the measured heat flow of the bacteria), X_{mp} (the error of the heat transfer fraction as shown above) and N_t (the error of the optical density measurement) are considered.

Table E.1: Overview over values and error during the methyl paraben experiment.

Value	**Intercept with y-axis (x=0)**	**Description**
Microcalorimetry 20 µL $\dot{Q}_0^{\mu/20\mu L}$	a=$4.6\times10^{-5}\pm0.4\times10^{-8}$ W	For 20 µL
Microcalorimetry 20 µL scaled $\dot{Q}_0^{\mu}$	$1.1\times10^{-4}\pm2.2\times10^{-7}$ W	For 46 µL
Sensor 1 methyl paraben $\dot{Q}_{m0}$	$8.3\times10^{-5}\pm8.8\times10^{-6}$ W	For 46 µL
Sensor 2 methyl paraben	$5.5\times10^{-5}\pm9.2\times10^{-6}$ W	-
Value	**Heat transfer fraction**	-
Sensor 1 X_{mp}	77 ± 8 %	-
Sensor 2 X_{mp}	52 ± 9 %	-

E.4 Parameter Overview over Microfluidic Chip

The Table E.2 is an overview over the main key performance metrics of the differential microfluidic calorimetric chip. Some of the values are derived from the system data analysis and some are from the lumped element model analysis.

Overview over Parameters		
Name	Value	Description
Limit of detection	1707 W/m^3	Limit of detection calculated in section 5.1.2.4
Standard deviation of noise	0.015 W/m^2, 15 nW	Standard deviation of noise
Sensitivity	0.73 $(W/m^2)/OD$	Sensitivity
Time constant τ	9 s	Lumped element model analysis calculated in section 3.4.1.1
Most important thermal conductivity k of PDMS	0.16 W/mK	Lumped element model analysis calculated in section 3.4.1.1
Thermal capacitance	0.25 J/K	Lumped element model analysis calculated in section 3.4.1.1
Heat transfer fraction X_{mp}	Modelled: 73% Experimental: 77% $\pm$ 8%	System data analysis elaborated on in section 3.4.1.1 and 5.1.2.2

Table E.2: Overview over relevant extracted parameters in the microfluidic chip.

E.5 Sensitivity, Resolution and the Limit of Detection

The standard deviation determined by the averaged heat flux data is 0.015 W/m^2.

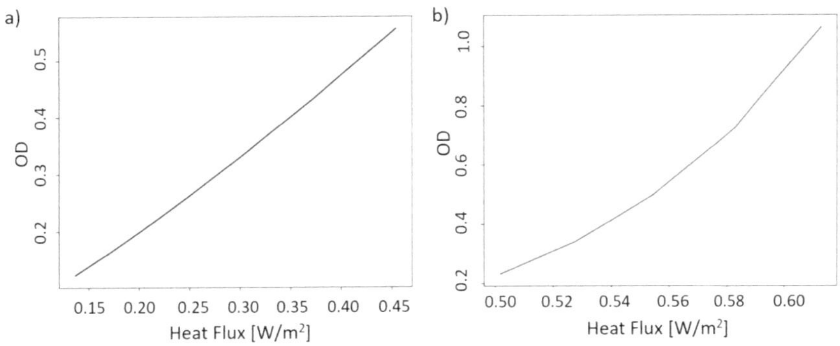

Figure E.4: Sensitivity (heat flux vs. optical density) of the two experiments. a) Data as shown in Figure 5.10 d). b) Data as shown in Figure 5.11 d). Open Access figure from [122] published under open access under CC BY license (Creative Commons Attribution 4.0 International license).

Figure E.4 shows the sensitivity of the different experiments as shown in Figure 5.10 and 5.11. In the first experiment, the sensitivity is $\frac{\partial N_t}{\partial \dot{Q}_e}$=1.36 OD/(W/m^2) and for the second experiment it is $\frac{\partial N_t}{\partial \dot{Q}_e}$=7.42 OD/(W/m^2). The limit of detection is defined as the standard deviation σ divided by the sensitivity. The calculated values are summarized in Table E.3.

Experiment	Standard Deviation	Sensitivity	Limit of Detection
Sensor 1	0.015 W/m^2	0.73 OD/(W/m^2)	0.021 OD
Sensor 1	1.52$\times$10^{-6} W	6.79$\times$10^{-14} W/(bacteria/mL)	22435882 bacteria/mL
Sensor 2	2.12$\times$10^{-6} W	0.13 OD/(W/m^2)	0.162 OD
Sensor 2	0.022 W/m^2	1.25$\times$10^{-14} W/(bacteria/mL)	174431495 bacteria/mL

Table E.3: Overview over the extracted standard deviation, sensitivity, and limit of detection in both experiments.

F Appendix - COMSOL Simulations

This part of the Appendix is based on a previous publication [1].

F.1 Fluid Dynamic Simulations of the Aorta

F.1.1 Constant Velocity Fitting Parameters a and c

A linear relationship between the velocity (0.35, 0.45, 0.55, and 0.65 m/s) of the fluid and the fitting parameters a and c as shown in Figure F.1. The fitting equation is $h = az^{-1/2} + cz^{-3/2}$ With this function, a predictability of the heat transfer coefficient in the low Reynolds number flow region in relation to the distance travelled from the inlet, z is found.

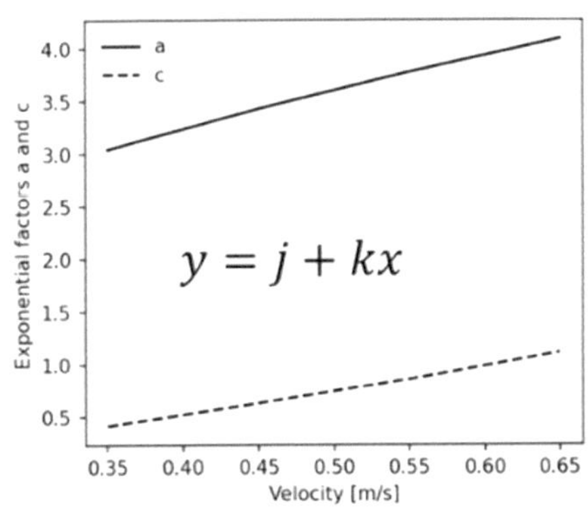

Figure F.1: Linear relationship between the fitting parameters a and c and the velocity of the fluid. Open Access figure adapted from [1] published under open access under CC BY license (Creative Commons Attribution 4.0 International license).

F.1.2 Full Flow Profiles

The full velocity profile in the radial direction of the different pulsatile flow conditions (parametric sweeps of A and f) is shown in Figure F.2.

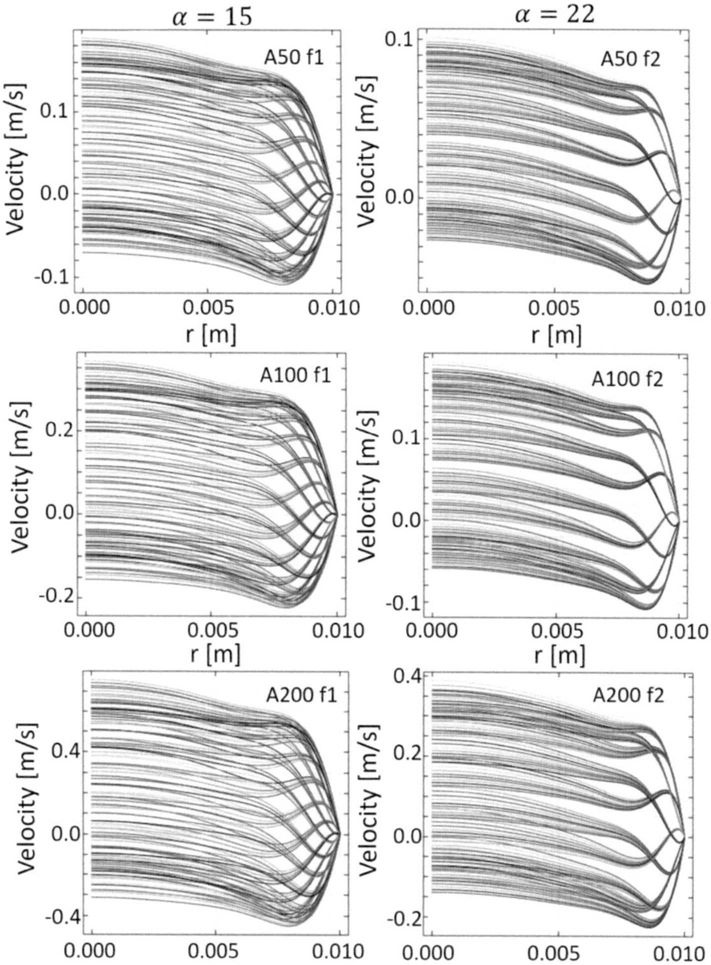

Figure F.2: Velocity profiles along the radial direction for the different swept parameters of A and f. α is the Womersley number. The color code shows different times over the time periods. Open Access figure adapted from [1] published under open access under CC BY license (Creative Commons Attribution 4.0 International license).

The heat transfer coefficients over the whole time range (10 s) of the simulation is extracted from the different parametric sweeps and shown in Figure F.3. Due to the transient behavior of the heat transfer coefficient over time, only the last 5 s are used for the averaged heat transfer coefficient values for the further data analysis.

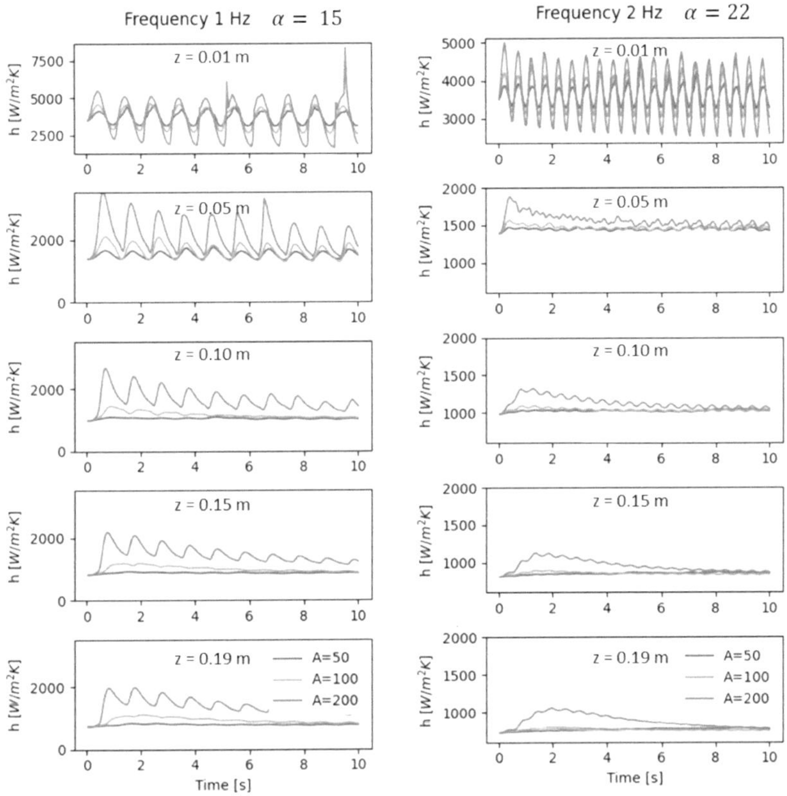

Figure F.3: Time dependent heat transfer coefficient along the aortic wall for the parametric sweeps. Open Access figure adapted from [1] published under open access under CC BY license (Creative Commons Attribution 4.0 International license).

F.1.3 Extracted Percentage Differences Between Pulsatile and Stationary Flow

Table F.1 and F.2 show the averaged heat transfer coefficients for sinusoidal flow (5 last seconds) and their respective percentage differences to the heat transfer coefficients at constant velocity. The percentage difference between the interpolated pressure profile and at constant velocity is shown in Figure F.3.

Table F.1: $f = 1$ Hz percentage difference between the pulsatile flow and the stationary velocity for the sinusoidal simulations.

Distance [m]	Amplitude [Pa/s]	h_{1Hz} [W/m^{2}K]	$h_{stationary}$ [W/m^{2}K]	Difference [%]
0.01	50	3622	3930	-8
	100	3598	4172	-14
	200	3610	4815	-25
0.05	50	1490	1592	-6
	100	1564	1660	-6
	200	1999	1812	10
0.10	50	1035	1146	-10
	100	1096	1181	-7
	200	1498	1256	20
0.15	50	857	956	-10
	100	930	980	-5
	200	1305	1028	27
0.19	50	765	863	-11
	100	855	863	-3
	200	1225	919	33

Table F.2: $f = 2$ Hz percentage difference between the pulsatile flow and the stationary velocity for the sinusoidal simulations.

Distance [m]	Amplitude [Pa/s]	h_{2Hz} [W/(m^{2}K)]	$h_{stationary}$ [W/(m^{2}K)]	Difference [%]
0.01	50	3626	3901	-7
	100	3640	4172	-9
	200	3574	4305	-17
0.05	50	1441	1555	-7
	100	1452	1610	-10
	200	1510	1691	-11
0.10	50	1024	1125	-9
	100	1022	1156	-12
	200	1074	1202	-11
0.15	50	861	941	-9
	100	842	964	-13
	200	899	997	-10
0.19	50	779	851	-8
	100	755	871	-13
	200	824	897	-8

Table F.3: Percentage difference between the interpolated pressure profile from an aortic pressure measurement [124] and the stationary velocity.

z	0.01 m	0.05 m	0.10 m	0.15 m	0.19 m
$h_{P_{int}}$	3562	1816	1298	1096	985
$h_{stationary}$	4432	1731	1231	1015	900
Difference [%]	-20	5	5	8	8

F.1.4 Radius Variation

The radius is varied for sinusoidal flow with an amplitude A = 100 Pa/s and a frequency f = 1 Hz. As shown in Figure F.4, an increasing radius increases the heat transfer coefficient.

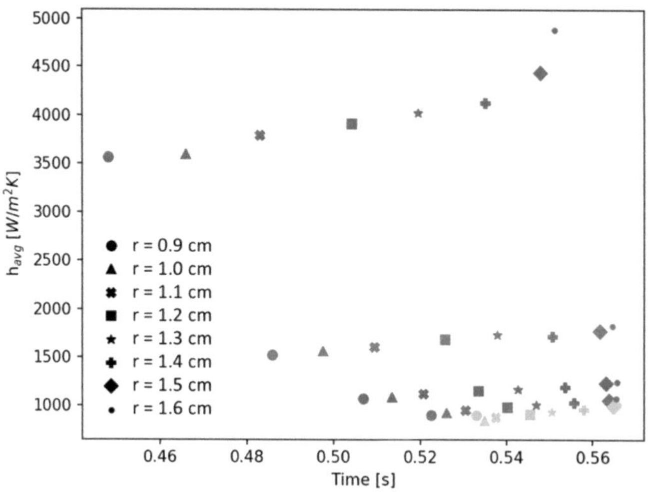

Figure F.4: Influence of radius variations on the heat transfer coefficient for pulsatile flow for an A = 100 Pa/s and f = 1 Hz at different distances from the inlet (red: 0.01 m, purple: 0.05 m, blue: 0.10 m, dark green: 0.15 m, light green: 0.19 m)

F.2 Thermodynamic Simulations of the Aorta

F.2.1 Thermodynamic Material Properties for the Vascular Graft

All the different polymeric materials used for the thermodynamic simulation of the aortic vascular graft materials. In addition, the thermal properties of aortic tissue are also presented.

Table F.4: Parameters used for the different simulation materials.

Property	PDMS	PTFE	ePTFE	Dacron (PETP)	Aortic tissue
k [W/mK]	0.16	0.2	0.35	0.51	0.46
c_p [J/kgK]	1460	960	2600	1300	960
ρ [kg/m^3]	970	1230	1230	1380	2200

F.2.2 Extracted Heat Transfer Efficiency for the Vascular Graft

The different percentage variations of the heat transfer efficiency of the aortic vascular graft upon changing the heat transfer coefficient is shown in Table F.5.

Table F.5: Difference in the heat transfer efficiency of the system between a heat transfer coefficient of h=200 W/(m^{2}K) and h=4800 W/(m^{2}K).

Material sweep heat transfer efficiency				
Material	Thickness	% at h=200 W/(m^{2}K)	% at h=4800 W/(m^{2}K)	% difference
PDMS	250 µm	43	66	23
PDMS	500 µm	35	47	12
PTFE	250µm	45	70	25
PTFE	500µm	38	533	15
ePTFE	250 µm	48	77	29
ePTFE	500µm	44	64	20
Dacron	250 µm	50	81	31
Dacron	500 µm	47	70	23
Aortic tissue	750 µm	42	59	17

F.2.3 Bacterial Populations and their Thermal Power Density

Table F.6: Thermal power densities and the respective amounts of bacteria in the case of an infection thickness of 450 µm and radius of 4 mm.

Heat transfer coefficient						
Thermal power density [W/m^3]	50	200	1000	2000	4000	10×10^3
#bacteria	3231	12.9×10^3	64.6×10^3	12.9×10^4	25.9×10^4	64.6×10^4
Comparison to biofilm [%]	0.01	0.06	0.29	0.57	1.14	2.86
Thermal power density [W/m^3]	50×10^3	10×10^4	20×10^4	30×10^4	35×10^4	-
#bacteria	32.3×10^5	6.46×10^5	12.9×10^6	19.4×10^6	22.6×10^6	-
Comparison to biofilm [%]	14.29	28.57	57.14	085.71	100.00	-

F.2.4 Increase in Infectious Thermal Power Densities

Figure F.5 shows increasing infectious thermal power densities in comparison to Figure 6.11.

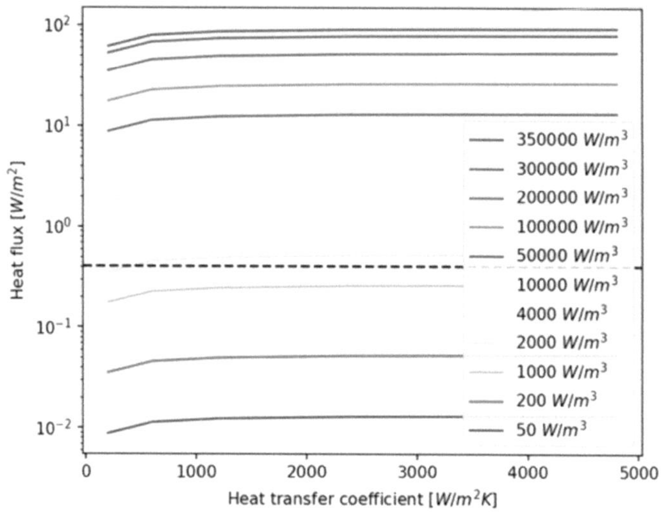

Figure F.5: Increasing the thermal power density of the infectious area for a radius of the infection of 4 mm, a thickness of 450 µm, and a location directly above the sensor ($d = 0$). Open Access figure from [1] published under open access under CC BY license (Creative Commons Attribution 4.0 International license).

F.2.5 Calculation of Distance Between Sensor and Infection

For the calculation of the distance between the sensor and the infection, the distance was varied using the arc angle, ϕ , which is defined between the middle of the sensor to the middle of the infection. The arc length, d, is defined as the distance between the middle of the infection and the middle of the sensor along the circumference of the arc as shown in Figure F.6.

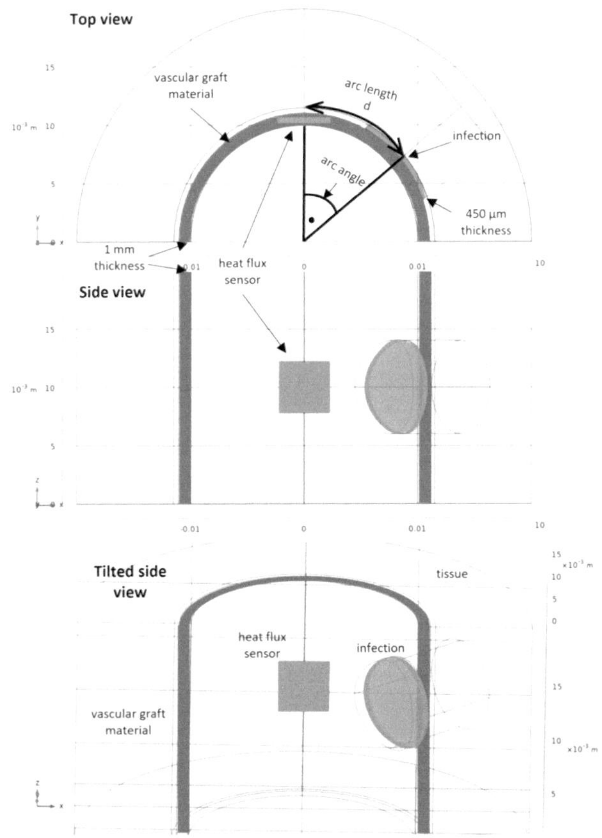

Figure F.6: The top view, side view, and tilted side view for the overview of the simulation build. Open Access figure from [1] published under open access under CC BY license (Creative Commons Attribution 4.0 International license).

The arc length, d, is defined as follows:

$$d = \phi \times \frac{\pi}{180} \times r \tag{F.1}$$

F.2.6 Temperature Increase at the Implant

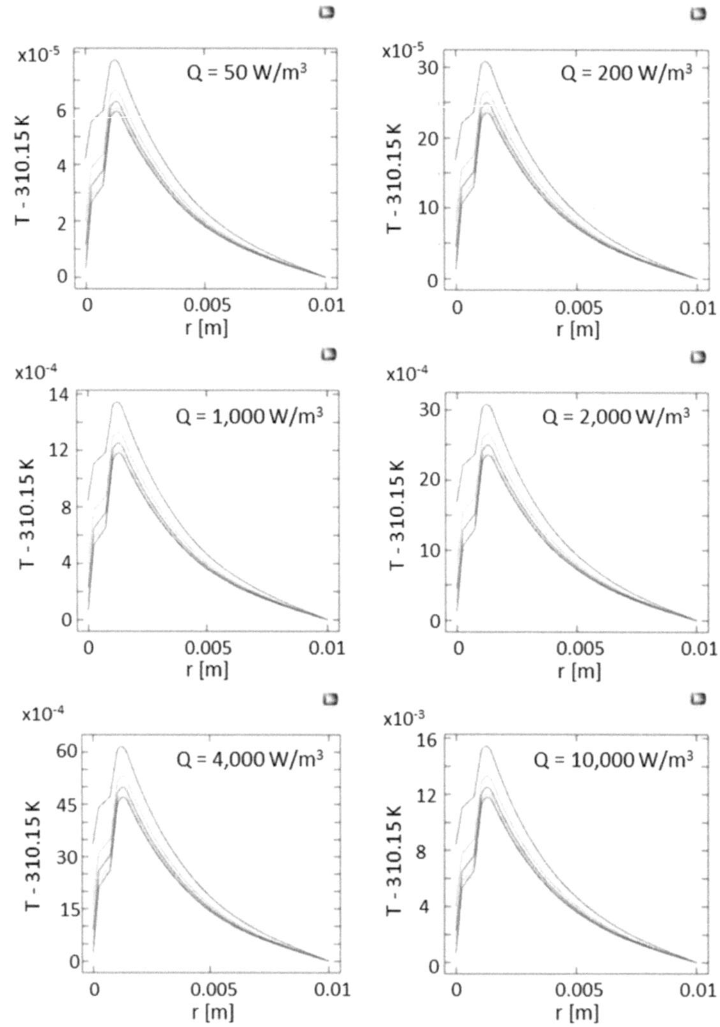

Figure F.7: Temperature profile through sensor, infection and tissue for different infectious thermal power density, Q, each with increasing heat transfer coefficients, h (200 W/(m^2K) (purple), 600 W/(m^2K) (light blue), 1200 W/(m^2K) (red), 2400 W/(m^2K) (green), and 4800 W/(m^2K) (dark blue) are represented by the different colors throughout the figures). Open Access figure from [1] published under open access under CC BY license (Creative Commons Attribution 4.0 International license).

F.2.7 Calculation of Mesh of Sensors

For the calculation of the hexagonal mesh of sensors we consider a 10 mm radius of a graft and a graft length, L, of 20 cm. The amount of sensors can be defined as the surface area of the cylinder divided by the unit cell area of the sensor as illustrated in Figure F.8:

$$N_{hexagonal} = \frac{graft\ surface\ area}{hexagonal\ unit\ cell\ area} = \frac{2\pi r L}{\frac{3\sqrt{3}}{2} d_{min}^2} = \frac{4\pi}{3\sqrt{3}} \frac{rL}{d_{min}^2} \tag{F.2}$$

Here, r is the radius of the graft, L is the graft length, and d_{min} as the minimum determined distance between the center of the sensor and the infection. According to this calculation, 177 sensors were calculated for $d_{min} = 5.24$ mm and 100 sensors for $d_{min} = 6.98$ mm.

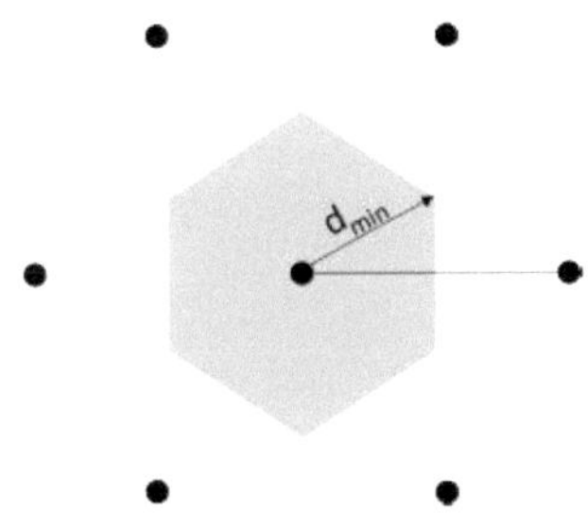

Figure F.8: The unit cell area of a hexagonal grid positioning of sensors. Each sensor is indicated by a dark circle. Open Access figure from [1] published under open access under CC BY license (Creative Commons Attribution 4.0 International license).

Bibliography

[1] S. L. K. Vehusheia, C. Roman, R. Sonderegger, N. Cesarovic, and C. Hierold, "Finite element model simulation of pulsatile blood flow towards the integration of heat flux sensors for localized early detection of aortic vascular graft infection," *Scientific Reports*, vol. 13, no. 16198, 2023, DOI: http://dx.doi.org/10.1038/s41598-023-42259-y.

[2] A. Voigt, A. Shalaby, and S. Saba, "Rising rates of cardiac rhythm management device infections in the united states: 1996 through 2003," *Journal of the American College of Cardiology*, vol. 48, no. 3, pp. 590–591, 2006, DOI: http://dx.doi.org/10.1016/j.jacc.2006.05.016.

[3] G. Herscu and S. E. Wilson, "Prosthetic infection: Lessons from treatment of the infected vascular graft," *Surgical Clinics of North America*, vol. 89, no. 2, pp. 391–401, 2008, DOI: http://dx.doi.org/10.1016/j.suc.2008.09.007.

[4] A. Nagpal and M. R. Sohail, "Prosthetic vascular graft infections: A contemporary approach to diagnosis and management," *Current Infectious Diseases Reports*, vol. 13, no. 4, pp. 317–323, 2011, DOI: http://dx.doi.org/10.1007/s11908-011-0191-y.

[5] C. R. Arciola, D. Campoccia, and L. Montanaro, "Implant infections: adhesion, biofilm formation and immune evasion," *Nature Reviews Microbiology*, vol. 16, no. 7, pp. 397–409, 2018, DOI: http://dx.doi.org/10.1038/s41579-018-0019-y.

[6] M. Jamal, F. M. Shaikh, B. Aslama, and K. M. Razeeb, "Sensor and biosensor to detect vascular graft infection: diagnosis and challenges," *Analytical Methods*, vol. 4, no. 7, pp. 1865–1875, 2012, DOI: http://dx.doi.org/10.1039/c2ay25027f.

[7] A. Celsus, *De Medicina*, First Century.

[8] H. J. Busscher, H. C. van der Mei, G. Subbiahdoss, P. C. Jutte, J. J. A. M. van den Dungen, S. A. J. Zaat, M. J. Schultz, and D. W. Grainger, "Biomaterial-associated infection: Locating the finish line in the race for the surface," *Science Translational Medicine*, vol. 4, no. 153, 2012, DOI: http://dx.doi.org/10.1126/scitranslmed.3004528.

[9] R. T. Southwood, J. L. Rice, P. J. McDonald, P. H. Hakendorf, and M. A. Rozenbilds, "Infection in experimental hip arthroplasties," *British Editorial Society of Bone and Joint Surgery*, vol. 67, no. 4, pp. 229–231, 1985, DOI: http://dx.doi.org/10.1302/0301-620X.67B2.3980532.

[10] M. Ribeiro, F. J. Monteiro, and M. P. Ferraz, "Infection of orthopedic implants with emphasis on bacterial adhesion process and techniques used in studying bacterial-material interactions," *Biomatter*, vol. 2, no. 4, pp. 176–194, 2012.

[11] M. Vert, Y. Doi, K.-H. Hellwich, M. Hess, P. Hodge, P. Kubisa, M. Rinauds, and F. Schue, "Terminology for biorelated polymers and applications," *Pure and Applied Chemistry*, vol. 84, no. 2, pp. 377–410, 2012, DOI: http://dx.doi.org/10.1351/PAC-REC-10-12-04.

[12] H. Wu, C. Moser, H.-Z. Wang, N. Hoeiby, and Z.-J. Song, "Strategies for combating bacterial biofilm infections," *International Journal of Oral Science*, vol. 7, 2015, DOI: http://dx.doi.org/10.1038/ijos.2014.65.

[13] D. Fleming and K. Rumbaugh, "The consequences of biofilm dispersal on the host," *Scientific Reports*, vol. 8, 2018, DOI: http://dx.doi.org/10.1038/s41598-018-29121-2.

[14] W. R. Wilson, T. C. Bower, M. A. Creager, S. Amin-Hanjani, P. T. OGara, P. B. Lockhart, R. O. Darouiche, B. Ramlawi, C. P. Derdeyn, A. F. Bolger, M. E. Levison, and K. A. Taubert, "Vascular graft infections, mycotic aneurysms, and endovascular infections," *AHA Scientific Statement*, 2016, DOI: http://dx.doi.org/10.1161/CIR.0000000000000457.

[15] C. Caradu, M. Puges, C. Cazanave, G. Martin, E. Ducasse, X. Berard, and C. Bicknell, "Outcomes of patients with aortic vascular graft and endograft infections initially contra-indicated for complete graft explantation," *Journal of Vascular Surgery*, vol. 1364-1373, no. 76(5), 2022, DOI: http://dx.doi.org/10.1016/j.jvs.2022.05.007.

[16] G. Zhou and T. Groth, "Host responses to biomaterials and anti-inflammatory design a brief review," *Macromolecular Bioscience*, vol. 18, no. 8, 2018, DOI: http://dx.doi.org/10.1002/mabi.201800112.

[17] A. Carnicer-Lombarte, S.-T. Chen, G. G. Malliaras, and D. G. Barone, "Foreign body reaction to implanted biomaterials and its impact in nerve neuroprosthetics," *Frontiers in Bioengineering and Biotechnology*, p. 271, 2021, DOI: http://dx.doi.org/10.3389/fbioe.2021.622524.

[18] P. Wojciechowski, P. T. Hove, and J. L. Brash, "Phenomenology and mechanism of the transient adsorption of fibrinogen from plasma (vroman effect)," *Journal of Colloid and Interface Science*, vol. 111, no. 2, pp. 455–465, 1986, DOI: http://dx.doi.org/10.1016/0021-9797(86)90048-2.

[19] E. F. Leonard and L. Vroman, "Is the vroman effect of impoartance in the interaction of blood with artificial materials?" *Journal of Biomaterials Science, Polymer Edition*, vol. 3, no. 1, pp. 95–107, 1992, DOI: http://dx.doi.org/10.1163/156856292X00105.

[20] L. Vroman, "Methods of investigating protein interactions on artificial and natural surfaces," *Annals of the New York Academy of Sciences*, vol. 516, no. 1, pp. 300–305, 1987, DOI: http://dx.doi.org/10.1111/j.1749-6632.1987.tb33050.x.

[21] K. Junge, M. Binnenboese, K. T. von Trotha, R. Rosch, U. Klinge, U. P. Neumann, and P. L. Janse, "Mesh biocompatibility: Effects of cellular inflammation and tissue remodelling," *Langenbeck's Archives of Surgery*, vol. 397, no. 2, pp. 255–270, 2012, DOI: http://dx.doi.org/10.1007/s00423-011-0780-0.

[22] J. M. Anderson, A. Rodriguez, and D. T. Chang, "Foreign body reaction to biomaterials," *Seminars in Immunology*, vol. 20, no. 2, pp. 86–100, 2008, DOI: http://dx.doi.org/10.1016/j.smim.2007.11.004.

[23] G. S. Selders, A. E. Fetz, M. Z. Radic, and G. L. Bowlin, "An overview of the role of neutrophils in innate immunity, inflammation and host-biomaterial integration," *Regenerative Biomaterials*, vol. 55-68, no. 4(1), 2017, DOI: http://dx.doi.org/10.1093/rb/rbw041.

[24] J. M. Anderson, A. G. Gristina, S. R. Hanson, L. A. Harker, R. J. Johnson, K. Merritt, P. T. Naylor, and F. J. Schoen, *Host Reactions to Biomaterials and Their Evaluation.* Elsevier Inc., 1996, DOI: http://dx.doi.org/10.1016/B978-0-08-050014-0.50009-2.

[25] Z. Sheikh, P. J. Brooks, O. Barzilay, N. Fine, and M. Glogauer, "Macrophages, foreign body giant cells and their response to implantable biomaterials," *Materials*, vol. 8, no. 9, pp. 5671–5701, 2015, DOI: http://dx.doi.org/10.3390/ma8095269.

[26] Q. Zhao, N. Topham, J. M. Anderson, A. Hiltner, G. Lodoen, and C. R. Payet, "Foreign-body giant cells and polyurethane biostability: In vivo correlation of cell adhesion and surface cracking," *Journal of Biomedical Materials Research*, vol. 25, no. 2, pp. 177–183, 1991, DOI: http://dx.doi.org/10.1002/jbm.820250205.

[27] S. Y. Chin, Y. C. Poh, A.-C. Kohler, J. T. Compton, L. L. Hsu, K. M. Lau, S. Kim, B. W. Lee, F. Y. Lee, and S. K. Sia, "Additive manufacturing of hydrogel-based materials for next-generation implantable medical devices," *Science Robotics*, vol. 2, pp. 1–2, 2017, DOI: http://dx.doi.org/10.1126/scirobotics.aah6451.

[28] W. K. Ward, "A review of the foreign-body response to subcutaneously-implanted devices: The role of macrophages and cytokines in biofouling and fibrosis," *Journal of diabetes science and technology*, vol. 6, no. 5, pp. 226–228, 2012, DOI: http://dx.doi.org/10.1177/193229680800200504.

[29] J. M. Anderson, "Biological responses to materials," *Annual Review of Materials Science*, vol. 31, pp. 81–110, 2001, DOI: http://dx.doi.org/10.1146/annurev.matsci.31.1.81.

[30] K. E. Piper, M. Fernandez-Sampedro, K. E. Steckelberg, J. N. Mandrekar, M. J. Karau, J. M. Steckelberg, E. F. Berbari, D. R. Osmon, A. D. Hanssen, D. G. Lewallen, R. H. Cofield, J. W. Sperling, J. Sanchez-Sotelo, P. M. Huddleston, M. B. Dekutoski, M. Yaszemski, B. Currier, and R. Patel, "C-reactive protein, erythrocyte sedimentation rate and orthopedic implant infection," *PLos ONE*, vol. 5, no. 2, pp. 5–9, 2010, DOI: http://dx.doi.org/10.1371/journal.pone.0009358.

[31] X. Ji, P. Zhou, L. Zhong, A. Xu, A. C. O. Tsang, and P. K. L. Chan, "Smart surgical catheter for c-reactive protein sensing based on an imperceptible organic transistor," *Advanced Science*, vol. 5, no. 6, pp. 1–8, 2018, DOI: http://dx.doi.org/10.1002/advs.201701053.

[32] M. H. Meyer, M. Hartmann, H.-J. Krause, G. Blankenstein, B. Mueller-Chorus, J. Oster, P. Miethe, and M. Keusgen, "Crp determination based on a novel magnetic biosensor," *Biosensors and Bioelectronics*, vol. 22, no. 6, pp. 973–979, 2007, DOI: http://dx.doi.org/10.1016/j.bios.2006.04.001.

[33] Y. Chen, B. Lu, Y. Chen, and X. Feng, "Breathable and stretchable temperature sensors inspired by skin," *Scientific Reports*, vol. 5, pp. 1–11, 2015, DOI: http://dx.doi.org/10.1038/srep11505.

[34] A. Ring, H. Sorg, A. Weltin, D. J. Tilkorn, J. Kieninger, G. Urban, and J. Hause, "In-vivo monitoring of infection via implantable microsensors: a pilot study," *Biomedizinische Technik*, vol. 63, no. 4, pp. 421–426, 2018, DOI: http://dx.doi.org/10.1515/bmt-2016-0250.

[35] N. J. Trengove, S. Langton, and M. C. Stacey, "Biochemical analysis of wound fluid from non healing and healing chronic leg ulcers," *Wound Repair and Regeneration*, vol. 4, no. 2, pp. 234–239, 1996, DOI: http://dx.doi.org/10.1046/j.1524-475X.1996.40211.x.

[36] P. Arquin, A. van den Berg, B. H. van der Schoot, N. F. de Rooij, H. Buehler, W. E. Morf, and L. F. J. Duerselen, "Integrated blood-gas sensor for po$_2$, pco$_2$, and ph," *Sensors and Actuators: B Chemical*, vol. 13, pp. 340–344, 1993, DOI: http://dx.doi.org/10.1016/0925-4005(93)85396-R.

[37] L. Santos, J. P. Neto, A. Crespo, D. Nunes, N. Costa, I. M. Fonseca, P. Barquinha, L. Pereira, J. Silva, R. Martins, and E. Fortunato, "Wo$_3$ nanoparticle-based conformable ph sensor," *ACS Applied Materials and Interfaces*, vol. 6, no. 15, pp. 12 226–12 234, 2014, DOI: http://dx.doi.org/10.1021/am501724h.

[38] P. Salvo, N. Calisi, B. Melai, V. Dini, C. Paoletti, T. Lomonaco, A. Pucci, F. D. Francesco, A. Piaggesi, and M. Romanelli, "Temperature- and ph-sensitive wearable materials for monitoring foot ulcers," *International Journal of Nanomedicine*, vol. 12, pp. 949–954, 2017, DOI: http://dx.doi.org/10.2147/IJN.S121726.

[39] M. Fierheller and R. G. Sibbald, "A clinical investigation into the relationship between increased periwound skin temperature and local wound infection in patients with chronic leg ulcers," *Advances in skin and wound care*, vol. 23, no. 8, pp. 369–379, 2010, DOI: http://dx.doi.org/10.1097/01.ASW.0000383197.28192.98.

[40] K. Roback, "An overview of temperature monitoring devices for early detection of diabetic foot disorders," *Expert Review of Medical Devices*, vol. 7, no. 5, pp. 711–718, 2010, DOI: http://dx.doi.org/10.1586/erd.10.35.

[41] M. C. Powanda and E. D. Moyer, "Plasma proteins and wound healing," *Surgery Gynecology and Obstetrics*, vol. 153, no. 5, pp. 749–755, 1981.

[42] U. Uzair, D. Benza, C. J. Behrend, and J. N. Anker, "Noninvasively imaging ph at the surface of implanted orthopedicdevices with x-ray excited luminescence chemical imaging," *AC Sensors*, vol. 4, no. 9, pp. 2367–2374, 2019, DOI: http://dx.doi.org/10.1021/acssensors.9b00962.

[43] D.-H. Kim, S. Wang, H. Keum, R. Ghaffari, Y.-S. Kim, H. Tao, B. Panilaitis, M. Li, Z. Kang, F. Omenetto, Y. Huang, and J. A. Rogers, "Thin, flexible sensors and actuators as instrumented surgical sutures for targeted wound monitoring and therapy," *Small*, vol. 8, no. 21, pp. 3263–3268, 2012, DOI: http://dx.doi.org/10.1002/smll.201200933.

[44] S. S. Karipott, P. M. Veetil, B. D. Nelson, R. E. Guldberg, and K. G. Ong, "An embedded wireless temperature sensor for orthopedic implants," *IEEE Sensors Journal*, vol. 18, no. 3, pp. 1265–1272, 2018, DOI: http://dx.doi.org/10.1109/JSEN.2017.2780226.

[45] J. M. Morais, F. Papadimitrakopoulos, and D. J. Burgess, "Biomaterials/tissue interactions: Possible solutions to overcome foreign body response," *The AAPS Journal*, vol. 12, no. 2, pp. 188–196, 2010, DOI: http://dx.doi.org/10.1208/s12248-010-9175-3.

[46] D. W. Grainger, "All charged up about implanted biomaterials," *Nature Biotechnology*, vol. 31, no. 6, pp. 507–509, 2013, DOI: http://dx.doi.org/10.1038/nbt.2600.

[47] Y. Zhang, R. C. Webb, H. Luo, Y. Xue, J. Kurniawan, N. H. Cho, S. Krishnan, Y. Li, Y. Huang, and J. A. Rogers, "Theoretical and experimental studies of epidermal heat flux sensors for measurements of core body temperature," *Advanced Healthcare Materials*, vol. 5, pp. 119–127, 2016, DOI: http://dx.doi.org/10.1002/adhm.201500110.

[48] S.-O. Yun, H. W. Cho, J.-h. Park, B. G. Choi, T. J. Lee, S. J. Kweon, J.-K. Lee, C.-H. Seo, Yoo, Hyung-Joun, and C.-Y. Kim, "Flexible ph sensor and system fabricated using pet film," *IEEE Sensors*, pp. 1–3, 2017, DOI: http://dx.doi.org/10.1109/ICSENS.2017.8233925.

[49] M. S. Wilson and W. Nie, "Multiplex measurement of seven tumor markers using an electrochemical protein chip," *Analytical Chemistry*, vol. 78, no. 18, pp. 6476–6483, 2006, DOI: http://dx.doi.org/10.1021/ac060843u.

[50] G. Gethin, "The significance of surface ph in chronic wounds," *Wounds UK*, vol. 3, pp. 52–56, 2007.

[51] P. S. Stewart and J. W. Costerton, "Antibiotic resistance of bacteria in biofilms," *Lancet*, vol. 358, pp. 135–138, 2001, DOI: http://dx.doi.org/10.1016/S0140-6736(01)05321-1.

[52] Y. K. Kim, E. Y. Chen, and W. F. Liu, "Biomolecular strategies to modulate the macrophage response to implanted materials," *Journal of Materials Chemistry B*, vol. 4, no. 9, pp. 1600–1609, 2016, DOI: http://dx.doi.org/10.1039/C5TB01605C.

[53] T. Maskow, R. Kemp, F. Buchholz, T. Schubert, B. Kiesel, and H. Harms, "What heat is telling us about microbial conversions in nature and technology: From chip- to megacalorimetry," *Microbial Biotechnology*, vol. 3, no. 3, pp. 269–284, 2010, DOI: http://dx.doi.org/10.1111/j.1751-7915.2009.00121.x.

[54] S. Wang, X. Sha, S. Yu, and Y. Zhao, "Nanocalorimeters for biomolecular analysis and cell metabolism monitoring," *Biomicrofluidics*, vol. 14, 2020, DOI: http://dx.doi.org/10.1063/1.5134870.

[55] B. B. Lowell and B. M. Spiegelman, "Towards a molecular understanding of adaptive thermogenesis," *Nature*, vol. 404, no. 6778, pp. 652–660, 2000, DOI: http://dx.doi.org/10.1038/35007527.

[56] D. C. Zaharia, A. A. Muntean, M. G. Popa, A. T. Steriade, O. Balint, R. Micut, C. Iftene, I. Tofolean, V. T. Popa, C. Baicus, M. A. Bogdan, and M. I. Popa, "Comparative analysis of staphylococcus aureus and escherichia coli microcalorimetric growth," *BMC Microbiology*, vol. 13, 2013, DOI: http://dx.doi.org/10.1186/1471-2180-13-171.

[57] N. Germic, Z. Frangez, S. Yousefi, and H.-U. Simon, "Regulation of the innate immune system by autophagy: neutrophils, eosinophils, mast cells, nk cells," *Cell Death and Differentiation*, vol. 703-714, no. 26(4), 2019, DOI: http://dx.doi.org/10.1038/s41418-019-0295-8.

[58] M. Si-Tahar, L. Touqui, and M. Chignard, "Innate immunity and inflammation - two facets of the same anti-infectious reaction," *Clinical and Experimental Immunology*, vol. 156, no. 2, pp. 194–198, 2009, DOI: http://dx.doi.org/10.1111/j.1365-2249.2009.03893.x.

[59] U. Weiss, G.; Schaible, "Macrophage defense mechanisms against intracellular bacteria," *Immunological Reviews*, vol. 182-203, no. 264(1), 2015, DOI: http://dx.doi.org/10.1111/imr.12266.

[60] W. Chanput, J. J. Mes, H. F. J. Savelkoul, and H. J. Wichers, "Characterization of polarized thp-1 macrophages and polarizing ability of lps and food compounds," *Food and Function*, vol. 4, no. 2, pp. 266–276, 2013, DOI: http://dx.doi.org/10.1039/C2FO30156C.

[61] Z. Heng, Z. Congyi, W. Cunxin, W. Jibin, G. Chaojian, L. Jie, and L. Yuwen, "Microcalorimetric study of virus infection the effects of hyperthermia and 1b recombinant homo interferon on the infection process of bhk-21 cells by foot and mouth disease virus," *Journal of Thermal Analysis and Calorimetry*, vol. 79, no. 1, pp. 45–50, 2005, DOI: http://dx.doi.org/10.1007/s10973-004-0560-y.

[62] K. Al-Hallak, M. K. Sarfraz, S. Azarmi, G. Kohan, W. H. Roa, and R. Loebenberg, "Microcalorimetric method to assess phagocytosis: Macrophage-nanoparticle interactions," *AAPS Journal*, vol. 13, no. 1, pp. 20–29, 2011, DOI: http://dx.doi.org/10.1208/s12248-010-9240-y.

[63] D. Gebreselassie, L. Ljunggren, and M. Monti, "Energetic study of human lymphocytes and the metabolic response to antigenic stimulation," *Thermochimica Acta*, vol. 251, pp. 177–186, 1995, DOI: http://dx.doi.org/10.1016/0040-6031(94)02069-Z.

[64] C. Murigande, S. Regenass, D. Wirz, A. U. Daniels, and A. Tyndall, "A comparison between (3h)-thymidine incorporation and isothermal microcalorimetry for the assessment of antigen-induced lymphocyte proliferation," *Immunological Investigations*, vol. 38, no. 1, pp. 67–75, 2009, DOI: http://dx.doi.org/10.1080/08820130802572160.

[65] O. Sharif, V. N. Bolshakov, S. Raines, P. Newham, and N. D. Perkins, "Transcriptional profiling of the lps induced nf-kb response in macrophages," *BMC Immunology*, vol. 8, 2007, DOI: http://dx.doi.org/10.1186/1471-2172-8-1.

[66] A. Mullen, C. E. Loscher, and H. M. Roche, "Anti-inflammatory effects of epa and dha are dependent upon time and dose-response elements associated with lps stimulation in thp-1-derived macrophages," *Journal of Nutritional Biochemistry*, vol. 21, no. 5, pp. 444–450, 2010, DOI: http://dx.doi.org/10.1016/j.jnutbio.2009.02.008.

[67] J. Auwerx, "The human leukemia cell line, thp-1 : A multifacetted model for the study of monocyte-macrophage differentiation," *Experientia*, vol. 47, 1991, DOI: http://dx.doi.org/10.1007/BF02041244.

[68] W. Chanput, J. J. mes, and H. J. Wichers, "Thp-1 cell line: An in vitro cell model for immune modulation approach," *International Immunopharmacology*, vol. 23, no. 1, pp. 37–45, 2014, DOI: http://dx.doi.org/10.1016/j.intimp.2014.08.002.

[69] A. Chanmugam, A. Bargava, and C. Herman, "Heat transfer model and quantitative analysis of deep tissue injury," *Proceedings of the ASME 2012 International Mechanical Engineering Congress Exposition*, 2012, DOI: http://dx.doi.org/10.1115/IMECE2012-88405.

[70] M. Ferreira and J. Yanagihara, "A transient three-dimensional heat transfer model of the human body," *International Communications in Heat and Mass Transfer*, vol. 36, no. 7, pp. 718–724, 2009, DOI: http://dx.doi.org/10.1016/j.icheatmasstransfer.2009.03.010.

[71] M. Rastgar-Jazi and F. Mohammadi, "Parameters sensitivity assessment and heat source localization using infrared imaging techniques," *BioMedical Engineering OnLine*, vol. 16, no. 1, pp. 1–18, 2017, DOI: http://dx.doi.org/10.1186/s12938-017-0403-2.

[72] O. Braissant, D. Wirz, B. Goepfert, and A. Daniels, "The heat is on: Rapid microcalorimetric detection of mycobacteria in culture," *Tuberculosis*, vol. 90, no. 1, pp. 57–59, 2010, DOI: http://dx.doi.org/10.1016/j.tube.2009.11.001.

[73] M. E. Butini, G. Abbandonato, C. D. Rienzo, A. Trampuz, and M. D. Luca, "Isothermal microcalorimetry detects the presence of persister cells in a staphylococcus aureus biofilm after vancomycin treatment," *Frontiers in Microbiology*, vol. 10, 2019, DOI: http://dx.doi.org/10.3389/fmicb.2019.00332.

[74] X.-M. Rong, Q.-Y. Huang, D.-H. Jiang, P. Cai, and W. Liang, "Isothermal microcalorimetry: A review of applications in soil and environmental sciences," *Pedosphere*, vol. 17, 2007, DOI: http://dx.doi.org/10.1016/S1002-0160(07)60019-8.

[75] C. Alklint, L. Wadsoe, and I. Sjoholm, "Accelerated storage and isothermal microcalorimetry as methods of predicting carrot juice shelf-life," *Journal of the Science of Food and Agriculture*, vol. 85, no. 2, 2005, DOI: http://dx.doi.org/10.1002/jsfa.1942.

[76] A. Trampuz, S. Salzmann, J. Antheaume, and A. U. Daniels, "Microcalorimetry: a novel method for detection of microbial contamination in platelet products," *Transfusion*, vol. 47, no. 9, pp. 1643–1650, 2007, DOI: http://dx.doi.org/10.1111/j.1537-2995.2007.01336.x.

[77] L. Naessberger, E. Jensen, M. Monti, and C.-H. Florin, "Microcalorimetric investigation of metabolism in rat hepatocytes cultured on microplates and in cell suspensions," *Biochimica et Biophysica Acta*, vol. 882, pp. 353–358, 1986, DOI: http://dx.doi.org/10.1016/0304-4165(86)90258-8.

[78] M. Monti, *Thermal and Energetic Studies of Cellular Biological Systems*. Elsevier Ltd., 1987, DOI: http://dx.doi.org/10.1016/C2013-0-06478-X.

[79] S. A. Thorrn, M. Monti, and B. Holman, "Heat conduction microcalorimetry of overall metabolism in rabbit alveolar macrophages in monolayers and in suspensions," *Biochimica et Biophysica Acta*, vol. 1033, pp. 305–310, 1990, DOI: http://dx.doi.org/10.1016/0304-4165(90)90138-M.

[80] B. A. Wagner, S. Venkataraman, and G. R. Buettner, "The rate of oxygen utilization by cells," *Free Radical Biology and Medicine*, vol. 51, no. 3, pp. 700–712, 2011, DOI: http://dx.doi.org/10.1016/j.freeradbiomed.2011.05.024.

[81] R. Kuse, S. Schuster, S. Schuebbe, H. Dix, and K. Hausmann, "Blood lymphocyte volumes and diameters in patients with chronic lymphocytic leukemia and normal controls," *Blut*, vol. 50, pp. 243–248, 1985, DOI: http://dx.doi.org/10.1007/BF00320301.

[82] F. Krombach, S. Munzing, A.-M. Allmeling, J. Gerlach, J. Behr, , and M. Doerger, "Cell size of alveolar macrophages : An interspecies comparison," *Environmental health perspectives*, vol. 105, pp. 1261–1263, 1997, DOI: http://dx.doi.org/10.1289/ehp.97105s51261.

[83] T. Maskow, J. Lerchner, M. Peitzsch, H. Harms, and G. Wolf, "Chip calorimetry for the monitoring of whole cell biotransformation," *Journal of Biotechnology*, vol. 122, no. 4, pp. 431–442, 2006, DOI: http://dx.doi.org/10.1016/j.jbiotec.2005.10.008.

[84] Y. Wang, H. Zhua, J. Feng, and P. Neuzil, "Recent advances of microcalorimetry for studying cellular metabolicheat heat," *Trends in Analytical Chemistry*, vol. 143, 2021, DOI: http://dx.doi.org/10.1016/j.trac.2021.116353.

[85] E. Kazura, B. R. Lubbers, E. Dawson, J. A. Phillips III, and F. Baudenbacher, "Nano-calorimetry based point of care biosensor for metabolic disease management," *Biomedical Microdevices*, vol. 19, 2017, DOI: http://dx.doi.org/10.1016/j.trac.2021.116353.

[86] J. Bae, J. Zheng, H. Zhang, P. J. Foster, and J. J. Needleman, Daniel J.and Vlassa, "A micromachined picocalorimeter sensor for liquid samples with application to chemical reactions and biochemistry," *Advanced Science*, vol. 8, no. 5, pp. 1–11, 2021, DOI: http://dx.doi.org/10.1002/advs.202003415.

[87] W. Lee, W. Fon, B. W. Axelrod, and M. L. Roukes, "High-sensitivity microfluidic calorimeters for biological and chemical applications," *Proceedings of the National Academy of Sciences of the United States of America*, vol. 106, no. 36, pp. 15 225–15 230, 2009, DOI: http://dx.doi.org/10.1073/pnas.0901447106.

[88] F. E. Torres, P. Kuhn, D. De Bruyker, A. G. Bell, M. V. Wolkin, E. Peeters, J. R. Williamson, G. B. Anderson, G. P. Schmitz, M. I. Recht, S. Schweizer, L. G. Scott, J. H. Ho, S. A. Elrod, P. G. Schultz, R. A. Lerner, and R. H. Bruce, "Enthalpy arrays," *Proceedings of the National Academy of Sciences of the United States of America*, vol. 101, no. 26, pp. 9517–9522, 2004, DOI: http://dx.doi.org/10.1073/pnas.0403573101.

[89] M. I. Recht, D. D. Bruyker, A. G. Bell, M. V. Wolkin, E. Peeters, G. B. Anderson, A. R. Kolatkar, M. W. Bern, P. Kuhn, R. H. Bruce, and F. E. Torres, "Enthalpy array analysis of enzymatic and binding reactions," *Proceedings of the National Academy of Sciences of the United States of America*, vol. 377, no. 1, pp. 33–39, 2008, DOI: http://dx.doi.org/10.1016/j.ab.2008.03.007.

[90] E. A. Johannessen, J. M. R. Weaver, P. H. Cobbold, and J. M. Cooper, "Heat conduction nanocalorimeter for pl-scale single cell measurements," *Applied Physics Letters*, vol. 80, no. 11, 2002, DOI: http://dx.doi.org/10.1063/1.1457532.

[91] J. Lerchner, A. Wolf, F. Buchholz, F. Mertens, T. Neu, H. Harms, and T. Maskow, "Miniaturized calorimetry - a new method for real-time biofilm activity analysis," *Journal of Microbiological Methods*, vol. 74, no. 2-3, pp. 74–81, 2008, DOI: http://dx.doi.org/10.1016/j.mimet.2008.04.004.

[92] F. Mariana, F. Buchholz, J. Lerchner, T. R. Neu, H. Harms, and T. Maskow, "Chip-calorimetric monitoring of biofilm eradication with antibiotics provides mechanistic information," *International Journal of Medical Microbiology*, vol. 303, no. 3, pp. 158–165, 2013, DOI: http://dx.doi.org/10.1016/j.ijmm.2012.12.009.

[93] T. Hartmann, N. Barros, A. Wolf, C. Siewert, P. Volpe, J. Schemberg, A. Grodrian, E. Kessler, F. Hänschke, F. Mertens *et al.*, "Thermopile chip based calorimeter for the study of aggregated biological samples in segmented flow," *Sensors and Actuators B: Chemical*, vol. 201, pp. 460–468, 2014, DOI: http://dx.doi.org/10.1016/j.snb.2014.05.024.

[94] B. Wang and Q. Lin, "Temperature-modulated differential scanning calorimetry in a mems device," *Sensors and Actuators, B: Chemical*, vol. 180, pp. 60–65, 2013, DOI: http://dx.doi.org/10.1016/j.snb.2012.02.044.

[95] R. Padovani, T. Lehnert, R. Trouillon, and M. A. M. Gijss, "Nanocalorimetric platform for accuratethermochemical studies in microliter volumes," *RSC Advances*, vol. 118, no. 5, pp. 97 133–97 142, 2015, DOI: http://dx.doi.org/10.1039/C5RA22248F.

[96] J. Lerchner, P. O. L. Volpe, C. Lanaro, K. Y. Fertrin, F. F. Costa, D. M. Albuquerque, F. Haenschke, and F. Mertens, “A chip calorimetry-based method for the real-time investigation of metabolic activity changes in human erythrocytes caused by cell sickling,” *Journal of Thermal Analysis and Calorimetry*, vol. 136, pp. 771–781, 2019, DOI: http://dx.doi.org/10.1007/s10973-018-7692-y.

[97] Y. Jia, B. Wang, Z. Zhang, and Q. Lin, “A polymer-based mems differential scanning calorimeter,” *Sensors and Actuators A: Physical*, vol. 231, pp. 1–7, 2015, DOI: http://dx.doi.org/10.1016/j.sna.2014.07.011.

[98] D. Hoces, M. Arnoldini, M. Diard, C. Loverdo, and E. Slack, “Growing, evolving and sticking in a flowing environment: understanding iga interactions with bacteria in the gut,” *Immunology*, vol. 159, no. 1, pp. 52–62, 2020, DOI: http://dx.doi.org/10.1111/imm.13156.

[99] A. P. Soleimany and S. N. Bhatia, “Activity-based diagnostics: An emergingparadigm for disease detection and monitoring,” *Trends in Molecular Medicine*, vol. 26, no. 5, pp. 450–468, 2020, DOI: http://dx.doi.org/10.1016/j.molmed.2020.01.013.

[100] S. Zelenin, J. Hansson, S. Ardabili, H. Ramachandraiah, H. Brismar, and A. Russom, “Microfluidic-based isolation of bacteria from whole blood for sepsis diagnostics,” *Trends in Molecular Medicine*, vol. 37, no. 4, pp. 825–830, 2015, DOI: http://dx.doi.org/10.1007/s10529-014-1734-8.

[101] G. Lambert, L. Estevez-Salmeron, S. Oh, D. Liao, B. M. Emerson, T. D. Tslty, and R. H. Austin, “An analogy between the evolution of drug resistance in bacterial communities and malignant tissues,” *Nature Reviews Cancer*, vol. 11, no. 5, pp. 375–382, 2011, DOI: http://dx.doi.org/10.1038/nrc3039.

[102] V. S. Shirure, Y. Bi, M. B. Curtis, A. Lezia, M. M. Goedegebuure, S. P. Goedegebuure, R. Aft, R. C. Fields, and S. C. George, “Tumor-on-a-chip platform to investigate progression and drug sensitivity in cell lines and patient-derived organoids,” *Lab Chip*, vol. 18, no. 23, pp. 3687–3703, 2018, DOI: http://dx.doi.org/10.1039/C8LC00596F.

[103] D. Huh, B. D. Matthews, A. Mammoto, M. Montoya-Zavala, H. Y. Hsin, and D. E. Ingber, “Reconstituting organ-level lung functions on a chip,” *Science*, pp. 1662–1668, 2010, DOI: http://dx.doi.org/10.1126/science.1188302.

[104] K.-J. Jang, A. P. Mehr, G. A. Hamilton, L. A. McPartlin, S. Chung, K.-Y. Suhd, and D. E. Ingber, "Human kidney proximal tubule-on-a-chip for drug transport and nephrotoxicity assessment," *Lab Chip*, vol. 5, no. 9, pp. 1119–1129, 2013, DOI: http://dx.doi.org/10.1039/c3ib40049b.

[105] W. Postek and P. Garstecki, "Droplet microfluidics for high-throughput analysis of antibiotic susceptibility in bacterial cells and populations," *ACS*, 2021, DOI: http://dx.doi.org/10.1021/acs.accounts.1c00729.

[106] T. Mulholland, M. McAllister, S. Patek, D. Flint, M. Underwood, A. Sim, J. Edwards, and M. Zagnoni, "Drug screening of biopsy-derived spheroids using a self-generated microfluidic concentration gradient," *Scientific Reports*, vol. 8, no. 1, 2018, DOI: http://dx.doi.org/10.1038/s41598-018-33055-0.

[107] D. Canfield, E. Kristensen, and B. Thamdrup, *Advances in Marine Biology*. Elsevier Inc., 2005, ISBN: 978-0-12-026147-5.

[108] Q. Ye, Y. Zhang, and J. Wei, "A comprehensive review of pulsating flow on heat transfer enhancement," *Applied Thermal Engineering*, vol. 196, 2021, DOI: http://dx.doi.org/10.1016/j.applthermaleng.2021.117275.

[109] H. Chattopadhyay, F. Durst, and S. Ray, "Analysis of heat transfer in simultaneously developing pulsating laminar flow in a pipe with constant wall temperature," *International Communications in Heat and Mass Transfer*, vol. 33, no. 4, pp. 475–481, 2006, DOI: http://dx.doi.org/10.1016/j.icheatmasstransfer.2005.12.008.

[110] S. Y. Kim, B. H. Kang, and J. M. Hyun, "Heat transfer in the thermally developing region of a pulsating channel flow," *International Journal of Heat and Mass Transfer*, vol. 36, no. 17, pp. 4257–4266, 1993, DOI: http://dx.doi.org/10.1016/0017-9310(93)90088-N.

[111] H. Hemida, M. Sabry, A. Abdel-Rahim, and H. Mansour, "Theoretical analysis of heat transfer in laminar pulsating flow," *International Journal of Heat and Mass Transfer*, vol. 45, no. 8, pp. 1767–1780, 2002, DOI: http://dx.doi.org/10.1016/S0017-9310(01)00274-5.

[112] J.-C. Yu, Z.-X. Li, and T. Zhao, "An analytical study of pulsating laminar heat convection in a circular tube with constant heat flux," *International Journal of Heat and Mass Transfer*, vol. 47, no. 24, pp. 5297–5301, 2004, DOI: http://dx.doi.org/10.1016/j.ijheatmasstransfer.2004.06.029.

[113] M. A. Habib, A. M. Attya, A. I. Eid, and A. Z. Aly, "Convective heat transfer characteristics of laminar pulsating pipe air flow," *Heat and Mass Transfer*, vol. 38, no. 3, pp. 221–232, 2002, DOI: http://dx.doi.org/10.1007/s002310100206.

[114] O. I. Craciunescu and S. T. Clegg, "Pulsatile blood flow effects ontemperature distribution andheat transfer in rigid vessels," *Transactions of the ASME*, vol. 123, no. 5, pp. 500–505, 2001, DOI: http://dx.doi.org/10.1115/1.1392318.

[115] A. Ostadfar, *Biofluid mechanics: Principles and applications.* Academic Press, 2016, ISBN: 978-0128024089.

[116] D. Obrist, *Biofluiddynamics Lecture Notes FS20*, March 2020.

[117] COMSOL, *Heat Transfer Model User's Guide*, 1998-2018.

[118] Y. A. Cengel and A. J. Ghajar, *Heat and Mass Transfer: Fundamentals and Applications*, 6th ed. McGraw Hill, 2020, ISBN-10: 1260922618.

[119] A. Mersmann, *Thermische Verfahrenstechnik.* John Wiley Sons, 1970, ISBN: 978-3-540-23648-1.

[120] Y. S. Muzychka and M. M. Yovanovich, "Laminar forced convection heat transfer in the combined entry region of non-circular ducts," *Journal of Heat Transfer*, vol. 126, no. 1, pp. 54–61, 2004, DOI: http://dx.doi.org/10.1115/1.1643752.

[121] F. P. Incropera, D. P. Dewitt, T. L. Bergman, and A. S. Lavine, *Fundamentals of Heat and Mass Transfer Sixth Edition.* John Wiley Sons, 2007, ISBN-13: *978-0-471-45728-2.

[122] S. L. K. Vehusheia, C. Roman, O. Braissant, M. Arnoldini, and C. Hierold, "Enabling direct microcalorimetric measurement of metabolic activity and exothermic reactions onto microfluidic platforms via heat flux sensor integration," *Nature Microsystems and Nanoengineering*, vol. 9, no. 1, pp. 1–12, 2023, DOI: http://dx.doi.org/10.1038/s41378-023-00525-z.

[123] S. Vehusheia, C. Roman, N. Cesarovic, and C. Hierold, "Microfluidic thermal model for early detection of infection on aortic grafts," *Technical Digest of 22nd International Conference on Solid State Sensors, Actuators and Microsystems (TRANSDUCERS) in Kyoto IEEJ*, no. 951-954, 25-29 June 2023.

[124] D. A. McDonald, "The relation of pulsatile pressure to flow in arteries," *Journal of Physiology*, vol. 127, pp. 533–553, 1955, DOI: http://dx.doi.org/10.1113/jphysiol.1955.sp005275.

[125] S. L. K. Vehusheia, C. I. Roman, M. Arnoldini, and C. Hierold, "Experimental in vitro microfluidic calorimetric chip data towards the early detection of infection on implant surfaces," *MDPI Sensors*, vol. 1019, no. 24(3), 2024, DOI: http://dx.doi.org/10.3390/s24031019.

[126] S. L. K. Vehusheia, C. Roman, M. Arnoldini, and C. Hierold, "Microcalorimetric measurement on a microfluidic chip in a thermally fluctuating environment," *MDPI Proceedings, Presented at Eurosensors Conference Lecce 10-13 September*, 2023.

[127] M. O'Neill, A. Beezer, G. Vine, R. Kemp, D. Olomolaiye, P. Volpec, and D. Oliveira, "Practical and theoretical consideration of flow-through microcalorimetry: determination of 'thermal volume' and its flow rate dependence," *Thermochimica Acta*, vol. 413, pp. 193–199, 2004, DOI: http://dx.doi.org/10.1016/j.tca.2003.09.021.

[128] E. Baxter, A. Graham, N. Re, I. Carr, J. Robinson, S. Mackie, and A. Morgan, "Standardized protocols for differentiation of thp-1 cells to macrophages with distinct m (ifnγ+ lps), m (il-4) and m (il-10) phenotypes," *Journal of immunological methods*, vol. 478, p. 112721, 2020, DOI: http://dx.doi.org/10.1016/j.jim.2019.112721.

[129] G. Sezonov, D. Joseleau-Petit, and R. D'Ari, "Escherichia coli physiology in luria-bertani broth," *Journal of Bacteriology*, vol. 189, pp. 8746–8749, 2007, DOI: http://dx.doi.org/10.1128/jb.01368-07.

[130] J. Higuera-Guisset, J. Rodriguez-Viejo, M. Chacon, F. Munoz, N. Vigues, and J. Ma, "Calorimetry of microbial growth using a thermopilebased microreactor," *Thermochimica Acta*, vol. 427, no. 1-2, pp. 187–191, 2005, DOI: http://dx.doi.org/10.1016/j.tca.2004.09.010.

[131] O. Braissant, D. Wirz, B. Goepfert, and D. A. U., "Biomedical use of isothermal microcalorimeters," *Sensors*, vol. 10, no. 10, pp. 9369–9383, 2010, DOI: http://dx.doi.org/10.3390/s101009369.

[132] Z. Zhang and S. Tadigadapa, "Calorimetric biosensors with integrated microfluidic channels," *Biosensors Bioelectronics*, vol. 19, pp. 1733–1743, 2004, DOI: http://dx.doi.org/10.1016/j.bios.2004.01.009.

[133] J. Lerchner, A. Wolf, and G. Wolf, "Recent developments in integrated circuit calorimetry," *Journal of Thermal Analysis and Calorimetry*, vol. 57, no. 1, pp. 241–251, 1999, DOI: http://dx.doi.org/10.1023/A:1010152517237.

[134] A. Sontheimer-Phelps, B. A. Hassell, and D. E. Ingber, "Modelling cancer in microfluidic human organ-on-chips," *Nature Reviews Cancer*, vol. 65-81, no. 19(2), 2019, DOI: http://dx.doi.org/10.1038/s41568-018-0104-6.

[135] S. D. Watt, H. S. Sidhu, A. C. McIntosh, and J. Brindley, "Chaotic flow in competitive exothermic-endothermic reaction systems," *Applied Mathematics Letter*, no. 115, 2021, DOI: http://dx.doi.org/10.1016/j.aml.2020.106960.

[136] M. Kashid, O. Detraz, M. S. Moya, I. Yuranov, P. Ptrechtl, J. Membrez, A. Renken, and L. Kiwi-Minsker, "Micro-batch reactor for catching intermediates and monitoring kinetics of rapid and exothermic homogeneous reactions," *Chemical Engineering Journal*, vol. 149-156, no. 214, 2013, DOI: http://dx.doi.org/10.1016/j.cej.2012.10.019.

[137] R. Phillips, J. Kondev, J. Theriot, and H. G. Garcia, *Physical Biology of the Cell.* Garland Science, Taylor Francis Group, LLC, 2013, ISBN: 978-0-8153-4450-6.

[138] N. Narisawa, S. Furukawa, H. Ogihara, and M. Zamasaki, "Estimation of the biofilm formation of escherichia coli k-12 by the cell number," *Journal of Bioscience and Bioengineering*, vol. 99, no. 1, 2005, DOI: http://dx.doi.org/10.1263/jbb.99.78.

[139] C. Arens, S.; Kraft, U. Sclegel, G. Printzen, S. M. Perren, and M. Hansis, "Susceptibility to local infection in biological internal fixation experimental study of open vs minimally invasive plate osteosynthesis in rabbits," *Arch Orthop Trauma Surg*, vol. 82-85, no. 119, 1999, DOI: http://dx.doi.org/10.1007/s004020050361.

[140] N. Harrasser, J. Gorkotte, A. Obermeier, S. Feihl, M. Straub, J. Slotta-Huspenia, R. von Eisenart-Rothe, W. Moser, P. Gruner, M. de Wild, Gp, H. Witzer, and R. Burgkart, "A new model of implant-related osteomyelitis in the metaphysis of rat tibiae," *BMC Musculoskeletal Disorder*, no. 17, 2016, DOI: http://dx.doi.org/10.1186/s12891-016-1005-z.

[141] M. Monzon, F. Garcia-Alvarez, A. Lacleriga, E. Gracia, J. Leiva, C. Oteiza, and A. Amorena, "A simple infection model using pre-colonized implants to reproduce rat chronic staphylococcus aureus osteomyelitis and study antibiotic treatment," *Journal of Orthopaedic Research*, vol. 820-826, no. 19, 2001, DOI: http://dx.doi.org/10.1016/S0736-0266(00)00076-0.

[142] I. T. Gabe, J. H. Gault, J. Ross, D. T. Mason, C. J. Mills, J. P. Schillingford, and E. Braunwald, "Measurement of instantaneous blood flow velocity and pressure in conscious man with a catheter-tip velocity probe," *Circulation*, vol. 5, 1969, DOI: http://dx.doi.org/10.1161/01.CIR.40.5.603.

[143] C. T. Dotter, D. J. Roberts, and I. Steinberg, "Aortic length: Angiocardiographic measurements," *Circulation*, 1950, DOI: http://dx.doi.org/10.1161/01.CIR.2.6.915.

[144] F. Cuomo, S. Roccabianca, D. Dillon-Murphy, N. Xiao, J. D. Humphrey, and C. A. Figueroa, "Effects of age-associated regional changes in aortic stiffness on human hemodynamics revealed by computational modeling," *PLoS ONE*, vol. 12, no. 3, pp. 1–21, 2017, DOI: http://dx.doi.org/10.1371/journal.pone.0173177.

[145] C.-Y. Liu, D. Chen, D. A. Bluemke, C. O. Wu, G. Teixido-Tura, A. Chugh, S. Vasu, J. A. C. Lima, and W. G. Hundley, "Evolution of aortic wall thickness and stiffness with atherosclerosis: Long-term follow up from the multi-ethic study of atherosclerosis," *Hypertension*, vol. 65, pp. 1015–1019, 2005, DOI: http://dx.doi.org/10.1161/HYPERTENSIONAHA.114.05080.

[146] B. Wang, Y. Jia, and Q. Lin, "A microfabrication-based approach to quantitative isothermal titration calorimetry," *Biosensors and Bioelectronics*, vol. 78, pp. 438–446, 2016, DOI: http://dx.doi.org/10.1016/j.bios.2015.11.069.

[147] K. Verhaegen, K. Baert, J. Simaels, and W. Van Driessche, "A high-throughput silicon microphysiometer," *Sensors and Acutators, A: Physical*, vol. 82, no. 1, pp. 186–190, 2000, DOI: http://dx.doi.org/10.1016/S0924-4247(99)00365-9.

[148] J. M. Anderson, *Chapter 4: Mechanisms of Inflammation Infection With Implanted Devices.* Cardiovascular Pathology, 1999, DOI: http://dx.doi.org/10.1016/1054-8807(93)90045-4.

[149] A. Trampuz and W. Zimmerli, "Prosthetic joint infections: update in diagnosis and treatment," *Swiss Med Wkly*, vol. 243-251, no. 135, 2005.

[150] M. Ribiero, F. J. Monteiro, and M. P. Ferraz, "Infection of orthopedic implants with emphasis on bacterial adhesion process and techniques used in studying bacterial-material interactions," *Biomatter*, vol. 176-194, no. 2(4), 2012, DOI: http://dx.doi.org/10.4161/biom.22905.

[151] B. Xu, K. Ooi, C. Mavriplis, and M. Zaghloul, "Evaluation of viscous dissipation in liquid flow in microchannels," *Journal of Micromechanics and Microengineering*, vol. 13, pp. 53–57, 2003, DOI: http://dx.doi.org/10.1088/0960-1317/13/1/308.

Publications

Peer-reviewed journal publications

1. S. L. K. Vehusheia, C. I. Roman, O. Braissant, M. Arnoldini, C. Hierold, "Enabling direct microcalorimetric measurement of metabolic activity and exothermic reactions onto microfluidic platforms via heat flux sensor integration", *Microsystems and Nanoengineering*, 9(1), 1-12, 2023
2. S. L. K. Vehusheia, C. I. Roman, O. Braissant, R. Sonderegger, N. Cesarovic, C. Hierold, "Heat flux sensors for localized early detection of vascular graft infection - Impact of pulsatile blood flow on heat transfer coefficients ", *Scientific Reports*, 13(1), 16198, 2023
3. S. L. K. Vehusheia, C. I. Roman, M. Arnoldini, C. Hierold, "Experimental in vitro microfluidic calorimetric chip data towards the early detection of infection on implant surfaces", MDPI Sensors, 24, 1019, 2023

Refereed conference proceedings

1. S. L. K. Vehusheia, C. I. Roman, N. Cesarovic, C. Hierold, "Microfluidic thermal model for early detection of infection in aortic grafts", Transducers, June 25-29, Kyoto, Japan
2. S. L. K. Vehusheia, C. I. Roman, M. Arnoldini, C. Hierold, "Microcalorimetric measurement on a microfluidic chip in a thermally fluctuating environment", MDPI Proceedings, Presented at Eurosensors Conference, September 10-13, Lecce, Italy

Conference Talks

1. S. L. K. Vehusheia, C. I. Roman, M. Arnoldini, C. Hierold, "Microcalorimetric measurement on a microfluidic chip in a thermally fluctuating environment", Eurosensors, September 10-13, Lecce, Italy

Posters

1. S. L. K. Vehusheia, C. I. Roman, N. Cesarovic, C. Hierold, "Microfluidic thermal model for early detection of infection in aortic grafts", Transducers, June 25-29, Kyoto, Japan
2. S. L. K. Vehusheia, C. I. Roman, O. Braissamt, N. Cesarovic, M. Arnoldini, C. Hierold, "Microfluidic *in vitro* experimental platform towards the early detection of infection on implant surfaces using heat flux", DIGIPREDICT Consortium, April 27, Lausanne, Switzerland

Student project supervised

1. **Rafael Sonderegger:** *Bachelor Thesis:* "Investigation of the effect of pulsatile blood flow on local heat transfer in aortic implants"
2. **Loris Togninalli:** *Bachelor Thesis:* "*In vitro* system analysis of two real life situation and optimization with lumped element model"
3. **Christian Thiago Peterhans:** *Bachelor Thesis:* "Heat flux and temperature data readout and analysis in python"
4. **Liza Bachmann:** *Bachelor Thesis* "Analysis of heat transfer efficiency in a microfluidic chip with hydrolysis of methyl paraben"
5. **Lavinia Recchioni** *Master Semester Project* "Conjugate heat transfer investigation of in vitro microfluidic chip for the detection of bacterial growth"
6. **Elias Haller** *Bachelor Thesis* "Experimental investigation of heat transfer coefficient in pulsatile flow of aortic grafts"

Curriculum Vitae

Personal Details

Name	Signe Lin Kuei Vehusheia
Birth	06. May 1995, Oslo, Norway
Citizenship	Norwegian

Education

04/2019 – 06/2024 **Micro and Nanosystems, ETH Zurich, Switzerland**

PhD Dissertation "Exploring heat flux as a marker for the early detection of infection"

09/2016 – 10/2018 **Department of Materials Scienences, ETH Zurich, Switzerland**

MSc in Materials Science

04/2018 – 10/2018 **Doyle Group Chemical Engineering, Massachusetts Institute of Technology, USA**

Master Thesis: "Tuning interaction potentials to modify the structure and rheology of nanoemulsion colloidal gels"

10/2013 – 09/2016 **Department of Materials Science, RWTH Aachen, Germany**

BSc in Materials Science, Bachelor Thesis: "Optimization of the layer stack for laser-induced amorphization of antimony thin films"

Work Experience

04/2019 – 04/2024 **Micro and Nanosystems, ETH Zurich, Switzerland**

Research, teaching assistant, chemical laboratory responsible and ETHZ internal first responder

04/2018 – 10/2018 **MIT Innovation Initiative/ETH Entrepreur Club, MIT, USA**

Team leader for the InCube project

Languages

Norwegian	Mother Tongue
English	Native Proficiency
German	Native Proficiency
French	Good
Mandarin	Colloquial

Scientific Reports on Micro and Nanosystems

edited by Prof. Dr. Christofer Hierold
ETH Zürich Micro and Nanosystems

Vol. 28: Verena Maiwald, **A Microelectromechanical Switch for Bandpass Vibration Detection.**
1st Edition 2018. XVI, 164 pages. € 64,00. ISBN 978-3-86628-614-6

Vol. 29: Michelle Müller, **Micromechanical broadband vibration amplitude-amplifier for microseismic and acoustic emission detection.**
1st Edition 2019. XVIII, 168 pages. € 64,00. ISBN 978-3-86628-627-6

Vol. 30: Silvan Marc Staufert, **Conformal Parylene-C Media Separating Membranes for Pressure Sensing Applications in Ventricular Assist Devices.**
1st Edition 2019. XVIII, 150 pages. € 64,00. ISBN 978-3-86628-636-8

Vol. 31: Lalit Kumar**, Energy Dissipation, Clamping and Motional Currents in Suspended Room Temperature Carbon Nanotube Resonators.**
1st Edition 2019. XXIV, 182 pages. € 64,00. ISBN 978-3-86628-650-4

Vol. 32: Sebastian Eberle**, Ultra-clean suspended carbon nanotube gas sensors - concept for large scale fabrication and sensor characterization.**
1st Edition 2019. XX, 180 pages. € 64,00. ISBN 978-3-86628-659-7

Vol. 33: Laura Vera Jenni, **Optimization of CNT Contacts in Suspended CNTFETs and Post Dry-Transfer Processing**.
1st Edition 2019. XXII, 170 pages. € 64,00. ISBN 978-3-86628-660-3

Vol. 34: Ian Mihailovic, **Ag/BiSe memristors for sensor data storage: a novel concept for zero-power sense-log devices.**
1st Edition 2022. XXII, 146 pages. € 64,00. ISBN 978-3-86628-769-3

Vol. 35: Stefan Nedelcu, **Energy efficient analog mixed-signal front ends for CNT-FET NO2 air-quality nanosensors.**
1st Edition 2022. XVI, 232 pages. € 64,00. ISBN 978-3-86628-779-2

Vol. 36: Johannes Weichart, **Artificial Fingertip with Embedded High Resolution Tactile Sensing.**
1st Edition 2023. (10), XX, 158 pages. € 64,00. ISBN 978-3-86628-802-7

Vol. 37: Katrina Klösel, **Multifunctional materials: exploiting the versatility of Bi_2Se_3 for multimodal sensing and zero power sensor systems.**
1st Edition 2024. (8), XX, 188 pages. € 64,00. ISBN 978-3-86628-816-4

Vol. 38: Matthias Alexander Dupuch, **Implantable Pressure Sensor Encapsulation for Ventricular Assist Device Control.**
1st Edition 2024. (8), XX, 168 pages. € 64,00. ISBN 978-3-86628-834-8

Hartung-Gorre Verlag, Konstanz http://www.hartung-gorre.de